SECURING OUR PLANET

How to Succeed
When Threats Are Too Risky
and There's Really No Defense

SECURING OUR PLANET

How to Succeed When Threats Are Too Risky and There's Really No Defense

An Ark Communications Institute book

edited by

Don Carlson and Craig Comstock

Published by Jeremy P. Tarcher, Inc.
Los Angeles

Distributed by St. Martin's Press
New York

Library of Congress Cataloging-in-Publication Data

Securing our planet.

1. Peace. I. Carlson, Don, 1932– . II. Comstock, Craig. III. Ark Communications Institute (U.S.)
JX1963.S37 1986 327.1'72 86–14418
ISBN 0–87477–407–1

Editor-in-chief
Ark Communications Institute
250 Lafayette Circle
Lafayette, CA 94549

Produced by Schuettge & Carleton
Composition by Classic Typography
Cover designed by Norman Ung
Manufactured in the United States of America
10 9 8 7 6 5 4 3 2 1
First Edition

CONTENTS

All introductions are by Craig Comstock

Questions about the arms race are expected more from liberal politicians, college professors, and Hollywood entertainers than from entrepreneurs or athletes. Don Carlson is among the exceptions. Winner of a triple athletic scholarship to Stanford, he went on to create, with partners, a company that manages several billion dollars in assets. Now in his fifties, Carlson is active in a business group that monitors the Pentagon and seeks to increase stability between the superpowers.

As founder and president of the Ark Communications Institute, he is trying to help, as he says, "build an ark as big as the earth." In order to lay the keel, we need to shake off both the illusion that nothing's wrong and the fatalism that says, "Don't bother to try, the challenge is too big for you." Carlson likes challenges. In the field of international relations he acknowledges that he plays as an amateur, but in looking at the world created by the pros, he observes, "Maybe it's time for the rest of us to lend a hand." When asked about national security, he replies that he supports it strongly, "for *all* nations."

Breaking the Trance

Don Carlson

WITH REGARD TO THE ARMS RACE AND OUR RELATIONS with the Soviet Union, we act as if we're in a trance. This book with its companion, *Citizen Summitry,** tells some ways to break the trance.

It was George F. Kennan, the historian and former Ambassador to Moscow who said that "the superpowers have proceeded with their nuclear buildup like victims of some sort of hypnotism, like men in a dream. . . . " Another historian, E.P. Thompson, recently used a similar image. "If we ask the partisans on either side what the Cold War is now about," he wrote, "they regard us with the glazed eyes of addicts." It's as if they know it's dangerous to go on, but desperately want another fix.

**Citizen Summitry*, edited by Don Carlson and Craig Comstock (Jeremy P. Tarcher, Inc./St. Martin's Press, 1986).

One of the most brilliant writers on the arms race, Thomas Powers, not long ago put Thompson's question to a hundred people, Americans and Soviets. He knew it was a hard question but he had in mind a story about the composer Stravinsky, who had written a new piece with a difficult violin passage. After weeks of rehearsal, the brilliant soloist apologized to the composer, saying the passage was so difficult that no violinist could possibly play it. "I understand that," said Stravinsky. "What I'm after is the sound of someone *trying* to play it." What struck Powers as he asked people about the reasons for the Cold War is that nobody on either side had even been trying to answer it—"their eyes were not exactly glazed, but they were certainly blank."

This is terryifying and it's remediable.

When I was in grammar school it was still obligatory to read Alfred Lord Tennyson's "The Charge of the Light Brigade":

Theirs not to make reply,
Theirs not to reason why,
Theirs but to do or die.
Into the Valley of Death rode the six hundred. . . .

On the part of the soldiers this could be called bravery; but with regard to the general it has been described as a blunder. In a sense we now are again near that valley, on a far grander scale, but instead of being troops under military discipline we are citizens free to oppose the system that has put us in mortal danger. That is the purpose of this book.

In one of his vivid and sometimes careless images, Winston Churchill declared in 1955 that, in the nuclear age, "safety will be the sturdy child of terror and survival the twin brother of annihilation." Presumably he meant the brother of the *threat* of annihilation. In any case, Churchill was suggesting that with both sides in possession of nuclear weapons, neither would dare to use them, an assumption that continues to guide the policy of the superpowers. As John F. Kennedy observed, however, "every man, woman, and child lives under a nuclear Sword of Damocles hanging by the slenderest of threads capable of being cut at any moment by accident, miscalculation, or madness." Since these three factors figured heavily in the outbreak of both world wars in this century, the situation is less than stable, despite the almost incredible multiplication of warheads on both sides.

What to do? An American long recognized as a masterful military leader and increasingly acknowledged as a resourceful politician said, near the end of his career, "I like to believe that people in the long run are going to do more to promote peace than governments." This was Dwight Eisenhower. He added his judgment that "people want peace so much that one of these days governments had better get out of their way and let them have it."

Leaders have been claiming that wars are always the fault of "the other side," a point made also by those on the other side. The whole system moves along as if ruled by fate. Yet, as Frederick Vinson argued, "wars are not 'acts of God,' they are caused by man, by man-made institutions, by the way in which man has organized his society." Vinson believed, further, that "what man has made, man can change."

During the "Reagan revolution" Americans learned how possible it is to change long-accepted aspects of our political policies. With some justification, those who tend toward the right of the political spectrum suggest that the historically liberal approach to domestic and foreign problem-solving has been based on unrestrained spending or, in more crude language, "throwing money at the problem." They have called this policy undisciplined and unrestrained, devoid of guidelines and limits. It has been built, they say, on a foundation of permissiveness and indulgence. Very harsh words, but most of us who are honest with ourselves would probably admit that many well-meaning government programs end up being bottomless financial pits that didn't accomplish their original purposes anyway and, in some cases, even compounded the very problems they were intended to solve. Isn't it time to acknowledge that money alone is not the answer to most problems?

Not surprisingly, however, these same critical adjectives hold true when applied to what is perhaps the only governmental program that has sustained long-term, bi-partisan support. I'm talking about, of course, our "blank check" military spending program. I sometimes think that "defense spending" would not have had the "sacred cow" financial treatment that almost rivals God and Motherhood if the "War Department" had not changed its name to the Department of Defense in 1947, not so many years before it lost the power to defend the U.S. against Soviet nuclear missiles.

It has also been asserted that many liberals have been naive about our need for an effective national defense and are sometimes almost oblivious to the "real world," especially to threats from abroad. Perhaps it is time to acknowledge that some loyal Americans probably haven't been totally realistic about our need for a defense that contributes to true security. Let us now all agree that we need a strong, realistic military capable of accomplishing our national objectives and go on from there.

Neither liberal nor conservative administrations have adequately approached critical issues relating to the quantity and quality of manpower and weaponry with adequate vision or fiscal controls. Such matters which relate so closely to life and death can no longer be addressed merely from a short-sighted or a tactical vantage point. Instead, we need to adopt a long-term realistic global perspective that includes (among myriad other factors) the terrifying reality of the increased nuclear proliferation into other countries of the world. Time is working against us. Theodore Roosevelt pointed out that "nine-tenths of wisdom consists in being wise in time."

Every question and every decision relating to new weapon systems should always contain a very careful analysis of what alteration, if any, this will create in the strategic relationship of the superpowers—both in the short and the long term. Why do we need it? Where will it take us? What will be the response of the other side? Will it move us towards greater stability or greater danger? Is the weapon redundant? Is it being built, as have been many in recent years, not for defensive reasons but, instead, primarily to gain political and psychological advantages? Is it being build primarily because we contracted to do it and there is no way to stop it? Does it exist because of the number of jobs that would be lost if the program were not initiated or an existing program were canceled?

In strategic terms, does the weapon really deter, or does it instead increase the possibility of having to enter an unthinkable war? Does the weapon system have a viable use in terms of an actual or potential conflict in the world? Even if the weapon could be potentially used, the basic question is, of course, could that possible conflict be resolved better and more satisfactorily for all parties through other than military means? We need to ponder these and a hundred other similar questions each time a new and more technologically advanced weapon is conceived of or comes off the research track, whether or not that weapon is labelled offensive or defensive.

In addition to the incomplete analysis brought to our defense needs by left and right, neither end of the spectrum has been able to incorporate a sense of economic wisdom into national security planning. Because of the sheer magnitude and size of military spending, we want to be sure (as with all other Federal outlays) that it is based on a coherent view of what we are really trying to achieve. Ultimately, the foundation of our security strategy must be built on a careful assessment of real risks that weigh the potential benefits and burdens of decisions against costs. All of this has to be guided by a commitment to developing lasting alternatives to *both* war and surrender.

A blank check given to the military-industrial complex does little to contribute to our security. It may in fact contribute to our insecurity, if only by successfully undermining our economy and by adding enormously to already staggering Federal deficits and to escalating interest payments on that cumulative debt. This growing deficit is being fueled by vastly accelerated Department of Defense spending. Considered in terms of the fiscal authority over which Congress has full discretion, military spending in 1985 alone exceeded more than half of the Federal budget and is moving upward rapidly. Current fiscal policies which require massive Federal borrowing are crowding out new private capital formation and pushing an already overvalued dollar to new heights, all of which is adversely affecting our international trade position. In the long run each of these effects will force interest rates to new plateaus, damaging small and large businesses, home buyers, farmers, and all of us as consumers. Most economists believe that these staggering and growing deficits will in time contribute to a resumption of the inflationary spiral with which we became all too familiar only a few years ago. Additionally, our current fiscal policies affect our economic image abroad because of demonstration of our lack of financial controls.

Spending money is clearly not all that is required to be successful in establishing a strong defense. If it were, Andre Maginot, the French

Defense Minister who created the historically famous but militarily disastrous "Maginot Line" on the French-German border prior to World War II, would have gone down in history as one of the world's great military strategists. Instead, because he focused on the short-sighted use of concrete and expensive military hardware instead of developing a strategy that was both realistic and visionary, his name has become synonymous with failure in defense planning. Paul Warnke, former Assistant Secretary of Defense, put it well when he declared, "It's a chump's game. You can spend additional billions and the best you can hope for is that you won't be much worse off in national security."

Since before the invention of nuclear weapons, some opponents of war have been trying to reduce its ghastiliness by "controlling" the number, type, and emplacement of weapons, and the rules for using them. Actually, this entire campaign, conducted by some of the most decent, ingenious and persistent statesmen, has accomplished very little, except to remind us of one type of danger.

Arms controllers apparently act in the spirit of Henry David Thoreau who said that "men have become the tools of their tools"—but in the case of the Cold War, the tools upon which citizens can act are not missiles waiting in their silos and submarines, but the governments that order and control the missiles. I believe that we can move toward peace by creating institutions that favor peace and by ending or altering the character of institutions that encourage war. This book offers some examples of how to take such steps.

Some people have elaborate, highly sophisticated explanations to defend their fatalistic view that nothing can be done. I am deeply concerned that important sectors of the U.S. elite have silently surrendered to the prospect of nuclear war. Years ago people wrote articles saying "if" nuclear war ever happened, such and such "would" be the consequence. Now I often see articles saying "when" war comes, the following "will" happen. Richard Garwin, a thoughtful scientist deeply involved in weapons policy, has estimated the chances of nuclear war by the end of the century at 50 percent. As I write, that is only 14 years from now. He may be right. But the question is what we can do to improve these odds.

If we become mesmerized by the danger, we will lose. As Ralph Waldo Emerson said, "skepticism is slow suicide"—he might have said *fatalism.* If we must be skeptical, let's carry it to the ultimate until we are so skeptical that we refuse to believe that making peace is impossible.

I agree that things look bad at the moment. The Administration has committed the U.S. to a twenty-year, trillion-dollar "Strategic Defense Initiative." Knowing the helplessness we all feel about nuclear rockets, President Reagan has projected a defense against them. There are a few problems. Nobody has shown that the system would work against *present* Soviet missiles, much less against whatever they would develop to foil it. It appears that the defense would cost more, per missile stopped, than it would cost the other side simply to build more missiles. Even a high rate of effectiveness would allow through an absolute number of warheads that would devastate the U.S. And finally, to the extent that the Soviets believed the system might work, or that we *believed* it might, they might be tempted to seize their last chance to attack.

The late Republican Senator Arthur Vandenberg, when truly agitated, used to denounce proposals by saying things such as "the end unattainable, the means hairbrained, and the cost staggering." This applies to the arms race in general, and with special intensity to Star Wars. But the "strategic defense initiative" is only the latest extension of what we and the Soviets have been doing for decades. It was Ogden Nash who said, "progress was all right once but it went on too long."

As a businessman I understand trends and recognize that if you are headed in a direction that leads you to a bad end you had better stop or change direction, not accelerate. In the words of Harold Willens, author of the *Trimtab Factor*, "if the Ford Motor Company had continued to manufacture the Edsel after the evidence was in that it wouldn't sell, a great industrial empire would have gone bankrupt." In business terms, there comes a time to "cut your losses." In the case of the Cold War, the goal should be to stop the arms race and begin the process of working toward cooperation rather than attempted domination. Albert Schweitzer said it so simply: "there is no practical problem existing between nations whose importance is in any proportion to the tremendous losses which must be expected in an atomic war."

It's time to break the trance. And I have to add that it's not only the national security experts who act as if they are hypnotized. In the peace movement, people are often as obsessed by the dangers of war as Cold Warriors are by the evil designs of the Soviet Union. In book after book, in documentary films, in speeches, we are told about the awesome effects of nuclear war, the danger that it may occur, and the provocative nature of various particular new weapons.

All this is necessary information. As Johnathan Schell, author of *The Fate of the Earth*, has argued, "it may be only by descending into this hell in the imagination now that we can hope to escape descending into it in reality at some later time." Making these points has required courage, hard intellectual work, and imagination. Now, however, thanks to Schell, Helen Caldicott, the producers of "The Day After," and many others, anyone with eyes to see or ears to hear knows that we're in terrible danger.

However, as one of the oldest mottoes in politics reminds us, "you can't fight something with nothing." The next step is to find sources of hope so that we are not paralyzed by fear and fatalism; and in order to do that, we need sketches of what peace will look like, analyses of its causes, ideas for specific actions. With its companion volume, that is what this book is about.

1 Setting Positive Mutual Examples

One summer evening a pilot friend and I were flying back after dark to Concord, California. The small airport there had already turned off its beacon lights. Looking for landmarks, I noticed a brightly lit perimeter fence surrounding some bunkers and realized it was the local nuclear weapons storage facility used by the Navy. It was comforting to see that any terrorists attempting to sneak across the fence would be well illuminated, but it was less comforting to be reminded that a city near my own is home to enough nuclear warheads to kill millions of people. As we flew past the storage facility and made our approach to the nearby airport, I thought about how many other nuclear bunkers must exist throughout the U.S. and the Soviet Union and on the soil of their various allies.[1]

I knew the weapons were in Concord to protect me, to allow me and my fellow citizens to continue to enjoy a magnificent "life-style." I knew that scintillating brilliance had gone into developing nuclear warheads, partly at the Livermore Laboratory not far south of Concord, and also into developing "doctrine" that would govern their possible use. I had been told that our nuclear weapons were fitted with "permissive action links" that would block any unauthorized use. Like others, I trusted that our national leaders were not so maladroit or ideologically rabid as deliberately to start a nuclear war, or casually to allow a peripheral conflict to escalate into one. Specialists have written that the Soviets, though bullies, are cautious.

So why worry? I worry because of Murphy, the one with a law named after him, or her. In one of many variants, Murphy's Law states that if something *can* go wrong, it eventually will. Like many Irish sayings, this is well framed for cautionary, if not scientific, purposes. You can't really disprove it. If you point to a system that appears to be working well, Murphy would surely say, "Just wait."

With regard to the nuclear system, that's what we're doing. We're assuming that two or more nuclear powers can continue threatening one another forever, each knowing that the other has

the power to destroy it within thirty minutes, and all with little prospect of protecting populations, even if a "Star Wars" system were eventually developed.[2]

According to a recent book by a former Strategic Air Command officer, our land-based nuclear missiles would be fired *before* incoming warheads land in the U.S.[3] This means that the time necessary to check the data, notify the President, doublecheck the data, confer, reach a decision, and get the U.S. missiles and bombers off the ground is shorter than many of us have spent deciding which video recorder to buy or what to serve at a dinner party.

In contrast, the Cuban missile crisis of October, 1962, was played out over thirteen days. It was not a sudden attack, but a covert attempt by the Soviets to construct nuclear missile bases "ninety miles from our shores." In his memoir of the crisis, Robert F. Kennedy stresses that it was this relatively lengthy period that allowed careful deliberation and the development of alternatives to an attack that might have started a war—or led to even more provocative Soviet countermoves elsewhere. Nonetheless, the pressure was terrible. "Each one of us was being asked to make a recommendation which, if wrong and if accepted, could mean the destruction of the human race."[4]

During this crisis, John F. Kennedy was, by all reports, amazingly sensitive to the danger of miscalculation, accident, and pugnacity. In part, this was his character; but his circumspection had recently been reaffirmed by reading Barbara Tuchman's book, *The Guns of August*—the story of how the major European powers, in 1914, had tumbled into a war that nobody quite intended, that caused horrifying damage, and that for five years nobody knew how to end. I have often thought that among contemporary writers who deserve the Nobel peace prize, a high place must go to Tuchman for telling this cautionary tale in a form accessible to at least one leader who might otherwise have acted less wisely. I'd nominate Murphy for the prize, too, but only a physicist could win for a formulation as succinct as his.

Something akin to the Cuban missile crisis could happen any day, and as systems become more complex, things have a way of get-

ting out of hand. On the level of technology, 1986 has brought such examples as the explosion of the Challenger shuttle on the U.S. side and of the Chernobyl nuclear reactor on the Soviet side. Everyone who works with computers has stories of "crashes," "bugs," and plain errors. We're now moving rapidly toward a space defense system that would have no more than two minutes to detect, analyze, and respond to a missile attack. Obviously, to the extent that it could work at all, it would have to be controlled by computers. There's no time for a human, whether elected or not, to elbow into the decision loop.

Can a serious person, with a knowledge of history, of organizational behavior, and of technological failure, reasonably assure us that the present system of mutual nuclear threats can continue indefinitely without leading to disaster? That is the question. If the answer is "No," as I believe it is, we have to press urgently for alternatives.

Arms control has been one response to this accident waiting to happen. Many deeply thoughtful, devoted people have worked over the four decades since World War II in order to control the arms race. Like Tuchman's book, arms control agreements and the process of negotiating them may have taken just enough edge off some explosive situations to prevent a disaster that would otherwise have occurred. But it's difficult to identify a very long list of major weapons which, if the agreements had *not* been negotiated, would otherwise have been built and created a significantly more dangerous situation.

Prior to Kenneth Adelman's appointment as director of the U.S. Arms Control and Disarmament Agency, he gave an interview in which, according to the reporter's notes, he called arms control negotiations a "sham" or possibly a "shame." In his confirmation hearing, some Senators dug out this alleged statement to suggest Adelman was unfit to take charge of arms control. A sham is something false or empty that purports to be genuine. If the main effect of ACDA is to provide support for a sham, perhaps the Agency ought not to exist; yet, in the view of the opponents of Adelman's appointment, arms control is necessary because without it our civilization is more likely to be wrecked.

Setting Positive Mutual Examples

If Adelman used the word "sham," he may have been guilty of a political offense far worse than inaccuracy—namely, indiscretion. The U.S. has had some negotiators who, with vision, persistence, and skill, have sought to rein in the wild horse of the arms race. But taking the U.S. side as a whole, including opponents of arms control in the Pentagon and the Senate and among businesspeople with military contracts, and ideological critics of *any* agreement with the Soviets, it is hard to argue that we have been more ready for radical changes than the other side.

And the style of the talks has been perfectly suited to their result—a tacit agreement between the U.S. and the Soviet Union not to agree on a basic restriction of the arms race.[5] Each side has tabled proposals designed to do three things—make it look peace-loving, thus mollifying its own public, its allies, and "world opinion"; allow it to do more or less whatever it wants; and subtly disadvantage the other side as substantially as possible. Since neither side is stupid, almost the only terms on which they can reach a compromise are to constrain no weapon that either deeply wants to build.

Why do the superpowers engage in this charade? Basically, it's because the public disbelieves the experts who keep assuring us that, with regard to nuclear weapons, everything's under control; and it's because allies are justifiably terrified of "tactical" nuclear warheads being exploded on or near their territory. Unlike many experts who speculate about war in the comfort and safety of think-tanks, the public continues to believe in Murphy's Law. Very simply, they sense that if we build an enormously destructive interactive system of nuclear threats, it will sooner or later go off.

If the chances for nuclear war are judged to be much above zero, the present system is clearly too dangerous *not* to change. Shortly before the Three Mile Island reactor malfunctioned and nearly exploded, a distinguished commission concluded that the statistical probability of a major malfunction in a U.S. nuclear power plant was one in many thousands of years, just as a NASA official put the chances of a space shuttle booster exploding at one in 100,000. If a power reactor on one side explodes, as Chernobyl did, the reactors on the other side are not programmed to follow suit.

Unfortunately, the same cannot be said of a missile attack. If we take this situation seriously, we will want to find, in Robert Fuller's phrase, "a better game than war."

Everybody wants security. Just as a thought-experiment, imagine a workshop at which U.S. and Soviet participants, and others, would design a global security system that protected everybody's basic interests. At the start, models of all existing weapons would be on the table, like toy soldiers. The workshop leader would gather up all the weapons and put them on the sidelines. Then the teams would explore what kind of relationship they could create and decide how many weapons, of what kinds, they wanted back. They would be able to negotiate: "If I stay at this level, what would you do?" and "Why go back to our old mistakes?—let's do such and such instead." Maybe the leader would arrange that neither team would know whether, in the workshop, they were acquiring weapons for themselves or for the other side. And maybe each team would combine Soviets and Americans. If the participants weren't satisfied with the outcome of a game, they would start again and continue until they were satisfied.

It would be preposterous to assume that, in this workshop, the teams would come up with anything resembling our present system.

Even the same cadre of leaders under whom the present system grew would make big changes, if they could start over. At least that's what some of them have said. For example, I once asked Kenneth Adelman, director of the Arms Control and Disarmament Agency, whether he knew of any occasion when the Soviets could usefully have said to us, "If you don't deploy such and such, we'll stop working on our version, too, and we'll both be much better off." He replied, "Probably in the history of MIRVing."[6] For those not familiar with acronymic verbs, to MIRV is to fit a missile with several warheads each able to be "delivered" to its own target. Many participants who were present at the creation of the MIRV, such as Henry Kissinger, felt later that an enormous opportunity had been missed. The opportunity was to resist "winning" a technological race, on the ground that the Soviets would soon catch up, and then we'd both suffer increased danger. The opportunity was lost, just as a similar opportunity is now being lost with regard to cruise missiles, a development that's technically as clever and practically as dubious.

Unable to ignore this unhappy history of surrender to the technological dynamic, some leaders have cast about for alternatives to what might be called the Geneva model of negotiations. Kenneth Adelman, for example, has called for "arms control without agreements"—offering as an example the U.S. decision in the 1960s to share with the Soviets the principle of the permissive action link, a coding system intended to prevent unauthorized use of nuclear warheads. The idea, of course, was that the U.S. has an interest in preventing the launch of a missile by an errant Soviet firing officer. The hint was taken. No bargaining was involved. The Soviets soon developed a similar system. One national security elite had taught its counterpart how to strengthen central control.

However, even Adelman sees limits to this approach. In the interview, I asked about "systems that warn of an attack," noting that "one repeatedly hears about antiquated Soviet computers and software, and ours haven't been performing flawlessly." The U.S. had helped the Soviets tighten their command and control over nuclear weapons: "Would the U.S. have a comparable interest in seeing that the Soviets aren't getting false alarms?" I have no inside information suggesting the Soviet computers have ever given false alerts, but we know for sure that *ours* have and that theirs are said to be less sophisticated. Would we have an interest in helping them? "Yes," Adelman replied, "if we could help them avoid false alarms in their military without giving them too much computer technology that would increase their threat. . . . " This reply points to a difficult dilemma, but given that the Soviets *already* have the technology to destroy us, what is it worth to us to help them discredit false alarms of a U.S. attack?

Apart from sharing certain capabilities with the Soviets, is there an approach to negotiation that would encourage each side to emulate positive steps by the other, rather than merely to solidify its bargaining "position"? In this section, a couple of alternatives are described. One is what Charles Osgood labels GRIT—graduated reciprocation in tension-reduction; the other is what Roger Fisher and William

Ury call, more pithily, "getting to yes." No method of negotiating can take us beyond the prevailing limits of political will and imagination, just as no technical means of communication can substitute for openness of heart and desire to learn.[7] But in the case of both communication and negotiation, some methods are better than others. Here are some examples.

—Craig Comstock

1. William M. Arkin and Richard W. Fieldhouse, *Nuclear Battlefields: Global Links in the Arms Race* (Ballinger, 1985).
2. Robert M. Bowman (Lt. Col., USAF, ret.), *Star Wars: A Defense Expert's Case Against the Strategic Defense Initiative* (Tarcher, 1986).
3. Bruce G. Blair, *Strategic Command and Control: Redefining the Nuclear Threat* (The Brookings Institution, 1986).
4. Robert F. Kennedy, *Thirteen Days: A Memoir of the Cuban Missile Crisis* (Norton, 1969).
5. For the maneuvering behind the environmental protection agreement to stop testing in the atmosphere, concluded in the early 1960s, see Glenn T. Seaborg, *Kennedy, Khrushchev, and the Test Ban* (University of California Press, 1981); for the period through the mid-1970s, Alva Myrdal, *The Game of Disarmament: How the United States and Russia Run the Arms Race* (Pantheon, 1976); and for the early 1980s, Strobe Talbott, *Deadly Gambits: The Reagan Administration and the Stalemate in Nuclear Arms Control* (Knopf, 1984).
6. Interview with Kenneth Adelman, conducted by Ark Communications Institute, May 14, 1986, Los Angeles.
7. On the subject of communication, see *Citizen Summitry: Keeping the Peace When It Matters Too Much to be Left to Politicians,* edited by Don Carlson and Craig Comstock (Tarcher, 1986).

In relationships marked by intense hostility and by vulnerability to enormous damage, nations often try to frighten the other side as much as possible, in order to "deter" it. Fear is a dynamic studied by psychologists (who sometimes also deal with one of its opposites, hope). Charles Osgood spent his professional life as a psychologist and is now professor emeritus at the University of Illinois in Champagne-Urbana.

His ideas about "graduated reciprocation in tension reduction" (GRIT) were already in the air for several years prior to John F. Kennedy's speech at American University in 1963, in which the President announced that the U.S. had stopped testing nuclear bombs in the atmosphere and invited the Soviet Union to reciprocate. Regardless of the extent to which advisers to the President had been directly influenced by Osgood's ideas, this speech and many of the actions that followed it are a textbook illustration of GRIT.

The Way GRIT Works

Charles E. Osgood

I LEARNED ABOUT HOW TO REDUCE INTERNATIONAL tensions by observing how our government deliberately escalated tensions in certain circumstances and then by realizing that similar techniques would work in reverse. For example, a strategy of calculated escalation has four salient features: first, the steps are unilaterally initiated (we did not negotiate with the North Vietnamese about increasing the tempo of our bombing or moving it closer to Hanoi; we just did it unilaterally); second, each step propels the opponent into reciprocating if he can, with more aggressive steps of his own (our development of multiple nuclear warheads propels the Soviets into analogous developments); third; such steps are necessarily graduated in nature—by the unpredictability of technological breakthroughs, by the limitations imposed by logistics, and by the oscillating level of perceived threat. Fourth, calculated escalation is obviously a tension-increasing process, the termination of which is a military resolution—victory, defeat, or (in our time) even mutual annihilation.

Now, if we change this last feature of calculated escalation—shift it from tension-*in*duction to tension-*re*duction—we have the essence of a calculated de-escalation strategy in conflict situations. Nation A devises patterns of small steps, well within its own limits of security, designed to reduce tensions and induce reciprocating steps from nation B. If such unilateral initiatives are persistently applied, and reciprocation is obtained, then the margin for risk-taking is widened and somewhat larger steps can be taken. Both sides, in effect, begin edging down the tension ladder, and both are moving—within what they perceive as reasonable limits of national security—toward a political rather than a military resolution. Needless to say, successful application of such a strategy assumes that both parties to a conflict have strong motives to get out of it—which, obviously, *should* be the case for both superpowers in this nuclear age.

The focus of my own long-term concern at the international level has been the rationalization of a strategy alternative whose technical name is Graduated and Reciprocated Initiatives in Tension-reduction. While doodling at a conference in the early 60's, I discovered that the initials of this mind-boggling phrase spelled GRIT—and, although I generally take a dim view of acronyms, this one was not only easy for people to remember but also suggested the kind of determination and patience required to apply it successfully.

One of the aims of GRIT is to reduce and control international tension levels. Another is to create an atmosphere of mutual trust within which negotiations of critical military and political issues can have a better chance of succeeding. In other words, GRIT is not a substitute for the more familiar process of negotiation, but rather a parallel process designed to enable a nation to take the initiative in a situation where a dangerous "balance" of mutual fear exists—and, to the degree successful, GRIT smooths the path of negotiation.

However, being unconventional in international affairs, the GRIT strategy is open to suspicion abroad and resistance at home. Therefore, it is necessary to spell out the ground rules under which this particular "game" should be played—to demonstrate how national security can be maintained during the process, how the likelihood of reciprocation can be maximized, and how the genuineness of initiations and reciprocations can be evaluated.

Rules for Maintaining Security

Rule 1: *Unilateral initiatives must not reduce a nation's capacity to inflict unacceptable nuclear retaliation should it be attacked at that level.* Nuclear capacity can serve rational foreign policy (a) if it is viewed not only as a deterrent but also as a security base from which to take limited risks in the direction of reducing tensions, (b) if the retaliatory, second-strike nature of the capacity is made explicit, and (c) if only the minimum capacity required for effective deterrence is maintained and the arms race damped. Needless to say, *none* of these conditions has been met to date by the two nuclear superpowers. Not only are nuclear weapons ambiguous as to initiation or retaliation, but both strategic and tactical weapons are redundantly deployed and in oversupply as far as capacity for graded response to aggression is concerned.

Rule 2: *Unilateral initiatives must not cripple a nation's capacity to meet conventional aggression with appropriately graded conventional response.* Conventional forces are the front-line of deterrence and they must be maintained at rough parity in regions of confrontation. But the absolute level at which the balance is maintained is variable. The general rule would be to initiate unilateral moves in the regions of least tension and gradually extend them to what were originally the most tense regions.

Rule 3: *Unilateral initiatives must be graduated in risk according to the degree of reciprocation obtained from an opponent.* This is the self-regulating characteristic of GRIT that keeps the process within reasonable limits of security. If bona fide reciprocations of appropriate magnitude are obtained, the magnitude and significance of subsequent steps can be increased; if not, then the process continues with steps of about the same magnitude of risk. The *relative* risk thus remains roughly constant throughout the process.

Rule 4: *Unilateral initiatives should be diversified in nature, both as to sphere of action and as to geographical locus of application.* The reason for diversification is two-fold: first, in maintaining security, diversification minimizes weakening one's position in any one sphere (for example, combat troops) or any one geographical locus (for example,

Berlin); second, in inducing reciprocation, diversification keeps applying the pressure of initiatives having a common tension-reducing intent (and hopefully effect), but does not "threaten" the opponent by pushing steadily in the same sphere or locus and thereby limiting his options in reciprocating.

Rules for Inducing Reciprocation

Rule 5: *Unilateral initiatives must be designed and communicated so as to emphasize a sincere intent to reduce tensions.* Escalation and deescalation strategies cannot be "mixed"—in the sense that military men talk about the "optimum mix" of weapon systems. The reason is psychological: reactions to *threats* (aggressive impulses) are incompatible with reactions to *promises* (conciliatory impulses); each strategy thus destroys the credibility of the other. It is therefore essential that a complete shift in basic policy be clearly signaled at the beginning. The top leadership of the initiating power must establish the right atmosphere by stating the overall nature of the new policy and by emphasizing its tension-reducing intent. To avoid "self-sabotage," it must be kept in mind that *all* of one government's actions with respect to another have the function of communicating intent. Control over de-escalation strategies must be just as tight and pervasive as control over war-waging strategies, if actions implying incompatible intents are not to intrude and disrupt the process.

Rule 6: *Unilateral initiatives should be publicly announced at some reasonable interval prior to their execution and identified as part of a deliberate policy of reducing tensions.* Prior announcements minimize the potentially unstabilizing effect of unilateral acts, and their identification with a total GRIT strategy helps shape the opponent's interpretation of them. However, the GRIT process cannot *begin* with a large, precipitate and potentially unstabilizing unilateral action. For example, then-Senator Mansfield's proposal in May, 1971—to cut by about half the U.S. forces permanently stationed in Europe, in one fell swoop—would have been likely to destabilize NATO/Pact relations, threaten our allies, and possibly encourage Soviet politico-military probes.

Rule 7: *Unilateral initiatives should include in their announcement explicit invitation to reciprocation in some form.* The purpose of this "rule" is to increase pressure on an opponent —by making it clear that reciprocation of appropriate form and magnitude is essential to the momentum of GRIT—and to bring to bear pressures of world opinion. However, *exactly* specifying the form or magnitude of reciprocation has several drawbacks: having the tone of a demand rather than an invitation, it carries an implied threat of retaliation if the demand is not met; furthermore, the specific reciprocation requested may be based on faulty perceptions of the other's situation, and *this* may be the reason for failure to get reciprocation. It is the occurrence of reciprocation *in any form*, yet having the same tension-reducing intent, that is critical. Speaking psychologically, the greatest conciliatory impact upon an opponent in a conflict situation is produced by his own, voluntary act of reciprocating. Such behaviour is incompatible with his Neanderthal beliefs about the unalterable hostility and aggressiveness of the initiator—and once he *has* committed a reciprocating action, all of the cognitive pressure is upon modifying these beliefs.

Rules for Demonstrating Good Faith

Rule 8: *Unilateral initiatives that have been announced must be executed on schedule regardless of any prior commitments to reciprocate by the opponent.* This is the best indication of the firmness and bona fide nature of one's own intent to reduce tensions. The control over what and how much is committed is the graduated nature of the process; at the time-point when each initiative is announced, the calculation has been made in terms of prior reciprocation history that this step can be taken within reasonable limits of security. Failure to execute an announced step, however, would be a clear sign of ambivalence in intent. This is particularly important in the early stages, when announced initiatives are liable to the charge of "propaganda."

Rule 9: *Unilateral initiatives should be continued over a considerable period, regardless of the degree or even absence of reciprocation.* Like the steady pounding on a nail, pressure toward reciprocating builds up as announced act follows announced act of a tension-reducing nature, even though the individual acts may be small in significance. It is this characteristic of GRIT which at once justifies the use of the acronym—and raises the hackles of most military men. But if a perceived threat equals capability times intent, the essence of this strategy *is* the calculated manipulation of the intent component of this equation. It is always difficult to "read" the intentions of an opponent in a conflict situation, and they are usually very complex. In such a situation, GRIT can be applied consistently to encourage conciliatory intents and interpretations at the expense of aggressive ones.

Rule 10: *Unilateral initiatives must be as unambiguous and as susceptible to verification as possible.* Although actions do speak louder than words, even overt deeds are liable to misinterpretation. Inviting opponent verification via direct, on-the-spot observation or via indirect media observation (for example, televising the act in question), along with requested reciprocation in the verification of *his* actions, is ideal—and what little might be lost in the way of secrecy by both sides may be more than made up in a reduced *need* for secrecy on both sides. However, nations in conflict have intense mutual suspicions and therefore place a heavy emphasis on mutual secrecy. The strategy of GRIT can be directly applied to this problem: particularly in the early stages, when the risk potentials are small, observers could be publicly invited to guarantee the verifiability of doing what was announced—and, although entirely *without* explicit insistence on reciprocation by the opponent, the implication would be strong, indeed. Initiatives whose face-validities are very high should be designed—for example, initial pullbacks of forces from border confrontations—and they can operate gradually to reduce suspicion and resistance to verification procedures. This should accelerate as the GRIT process continues.

Some Applications of GRIT Strategy

Over the past 15 years or so there has been considerable experimentation with the GRIT strategy—but mostly in the psychological laboratory. There have been sporadic GRIT-like moves in the real world—for example, in the early 1960s the graduated and reciprocated pullback of U.S. and Soviet tanks, which were lined up practically snout-to-snout at the height of the Berlin Crisis—but for the most part in recent history these have been one-shot affairs, always tentatively made, and never reflecting any genuine change in basic strategy.

The one exception to this dictum was "The Kennedy Experiment," as documented in a significant 1967 paper by Amitai Etzioni. This real-world test of a strategy of calculated de-escalation was conducted in the period from June to November 1963. The first step was President Kennedy's speech at The American University on June 10, in which he outlined what he called "A Strategy of Peace," praised the Russians for their accomplishments, noted that "our problems are man-made . . . and can be solved by man," and then announced the first unilateral initiative—the United States was stopping all nuclear tests in the atmosphere and would not resume them unless another country did. Kennedy's speech was published in full in both *Izvestia* and *Pravda*, with a combined circulation of 10,000,000. On June 15, Premier Khrushchev reciprocated with a speech welcoming the U.S. initiative, and he announced that he had ordered production of strategic bombers to be halted.

The next step was a symbolic reduction in the trade barriers between East and West; on October 9, President Kennedy approved the sale of $250 million worth of wheat to the Soviet Union. Although the U.S. had proposed a direct America-Russia communication link (the "hot line") in 1962, it wasn't until June 20, 1963—after the Kennedy Experiment had begun—that the Soviets agreed to this measure. Conclusion of a test-ban treaty, long stalled, was apparently the main goal of the experiment: multilateral negotiations began in earnest in July, and on August 5, 1963, the test-ban treaty was signed. The Kennedy Experiment slowed down with deepening involvement in Vietnam—and came to an abrupt end in Dallas, Texas.

Had this real-world experiment in calculated de-escalation been a success? To most of the initiatives taken by either side, the other reciprocated, and the reciprocations were roughly proportional in significance. What about psychological impact? I do not think that anyone who lived through that period will deny that there was a definite warming of American attitudes toward Russians, and the same is reported for Russian attitudes toward Americans—they even coined their own name for the new strategy, "the policy of mutual example."

Presidential speeches rarely qualify as great reading. Despite such masterpieces as Lincoln's address at Gettysburg, some of Franklin Roosevelt's speeches, and Eisenhower's advice while saying farewell to the Presidency, the form often slips into empty rhetoric. On such occasions, either the thought is unclear, false, trite, or irrelevant, the style is slack or insincere, the performance is detached from any action, or all of the above adjectives may apply.

"A Strategy of Peace" stands among the great Presidential speeches. It put the country's foreign policy in a strikingly new context. It was accompanied by a major action—namely, the halting of nuclear tests in the atmosphere. While recalling our quarrel with the Soviet Union, the speech also expressed compassion for that country and extended an invitation to it. And the words spoken by Kennedy have an elegance that the ancient Greek historian Thucydides would have recognized. In short, the American University speech repays close attention. It's as relevant now as when it was delivered.

In June 1986, without acknowledgement, President Reagan borrowed a passage from this speech, about how Soviet losses in World War II were equivalent to the destruction of America east of Chicago. It remains to be seen whether Reagan will, as Kennedy did, accompany these words with action.

A Strategy of Peace

John F. Kennedy

I HAVE CHOSEN THIS TIME AND PLACE TO DISCUSS A topic on which ignorance too often abounds and the truth is too rarely perceived—and that is the most important topic on earth: peace.

What kind of peace do I mean and what kind of peace do we seek? Not a *Pax Americana* enforced on the world by American weapons of war. Not the peace of the grave or the security of the slave. I am talking about genuine peace—the kind of peace that makes life on earth worth living—and the kind that enables men and nations to grow and to hope and build a better life for their children—not merely peace for Americans but peace for all men and women—not merely peace in our time but peace in all time.

Setting Positive Mutual Examples

I speak of peace because of the new face of war. Total war makes no sense in an age where great powers can maintain large and relatively invulnerable nuclear forces and refuse to surrender without resort to those forces. It makes no sense in an age when a single nuclear weapon contains almost ten times the explosive force delivered by all the Allied air forces in the Second World War. It makes no sense in an age when the deadly poisons produced by a nuclear exchange would be carried by wind and water and soil and seed to the far corners of the globe and to generations yet unborn.

Today the expenditure of billions of dollars every year on weapons acquired for the purpose of making sure we never need them is essential to the keeping of peace. But surely the acquisition of such idle stockpiles—which can only destroy and can never create—is not the only, much less the most efficient, means of assuring peace.

I speak of peace, therefore, as the necessary rational end of rational men. I realize the pursuit of peace is not as dramatic as the pursuit of war—and frequently the words of the pursuer fall on deaf ears. But we have no more urgent task.

Some say that it is useless to speak of peace or world law or world disarmament—and that it will be useless until the leaders of the Soviet Union adopt a more enlightened attitude. I hope they do. I believe we can help them do it.

But I also believe that we must re-examine our own attitudes—as individuals and as a nation—for our attitude is as essential as theirs. And every graduate of this school, every thoughtful citizen who despairs of war and wishes to bring peace, should begin by looking inward—by examining his own attitude towards the course of the cold war and towards freedom and peace here at home.

First: Examine our attitude towards peace itself. Too many of us think it is unreal. But that is a dangerous, defeatist belief. It leads to the conclusion that war is inevitable—that mankind is doomed—that we are gripped by forces we cannot control.

We need not accept that view. Our problems are man-made. Therefore, they can be solved by man. And man can be as big as he wants. No problem of human destiny is beyond human beings. Man's reason and spirit have often solved the seemingly unsolvable—and we believe they can do it again.

I am not referring to the absolute, infinite concepts of universal peace and goodwill of which some . . . fanatics dream. I do not deny the value of hopes and dreams, but we merely invite discouragement and incredulity by making that our only and immediate goal.

Let us focus instead on a more practical, more attainable peace—based not on a sudden revolution in human nature but on a gradual evolution in human institutions—on a series of concrete actions and effective agreement which are in the interests of all concerned.

There is no single, simple key to this peace—no grand or magic formula to be adopted by one or two powers. Genuine peace must be the product of many nations, the sum of many acts. It must be dynamic, not static, changing to meet the challenge of each new generation. For peace is a process—a way of solving problems.

With such a peace, there will still be quarrels and conflicting interests, as there are within families and nations. World peace, like community peace, does not require that each man love his neighbor—it requires only that they live together with mutual tolerance, submitting their disputes to a just and peaceful settlement. And history teaches us that enmities between nations, as between individuals, do not last forever. However fixed our likes and dislikes may seem, the tide of time and events will often bring surprising changes in the relations between nations and neighbors.

So let us persevere. Peace need not be impracticable—and war need not be inevitable. By defining our goal more clearly—by making it seem more manageable and less remote—we can help all people to see it, to draw hope from it, and to move irresistibly towards it.

And second: let us re-examine our attitude towards the Soviet Union. It is discouraging to think that their leaders may actually believe what their propagandists write.

It is discouraging to read a recent authoritative Soviet text on military strategy and find, on page after page, wholly baseless and incredible claims—such as the allegation that "American imperialist circles are preparing to unleash different types of war . . . that there is a very real threat of a preventative war being unleashed by American imperialists against the Soviet Union . . . [and that] the political aims," and I quote, "of the American imperialists are to enslave economically and politically the European and other capitalist countries . . . [and] to achieve world domination . . . by means of aggressive war."

Truly, as it was written long ago, "The wicked flee when no man pursueth." Yet it is sad to read these Soviet statements—to realize the extent of the gulf between us. But it is also a warning—a warning to the American people not to fall into the same trap as the Soviets, not to see only a distorted and desperate view of the other side, not to see conflict as inevitable, accommodation as impossible and communication as nothing more than an exchange of threats.

No government or social system is so evil that its people must be considered as lacking in virtue. As Americans, we find Communism profoundly repugnant as a negation of personal freedom and dignity. But we can still hail the Russian people for their many achievements—in science and space, in economic and industrial growth, in culture, and in acts of courage.

Among the many traits the peoples of our two countries have in common, none is stronger than our mutual abhorrence of war. Almost unique among the major world powers, we have never been at war with each other. And no nation in the history of battle ever suffered more than the Soviet Union in the Second World War. At least 20,000,000 lost their lives. Countless millions of homes and families were burned or sacked. A third of the nation's territory, including two-thirds of its industrial base, was turned into a wasteland—a loss equivalent to the destruction of this country east of Chicago.

Today, should total war ever break out again—no matter how—our two countries will be the primary targets. It is an ironic but accurate fact that the two strongest powers are the two in the most danger of devastation. All we have built, all we have worked for, would be destroyed in the first twenty-four hours. And even in the cold war—which brings burdens and dangers to so many countries, including this nation's closest allies—our two countries bear the heaviest burdens. For we are both devoting massive sums of money to weapons that could be better devoted to combat ignorance, poverty and disease.

We are both caught up in a vicious and dangerous cycle with suspicion on one side breeding suspicion on the other, and new weapons begetting counter-weapons.

In short, both the United States and its allies, and the Soviet Union and its allies, have a mutually deep interest in a just and genuine peace and in halting the arms race. Agreements to this end are in the interests of the Soviet Union as well as ours—and even the most hostile nation can be relied upon to accept and keep those treaty obligations, and only those treaty obligations, which are in their own interest.

So, let us not be blind to our differences—but let us also direct attention to our common interests and the means by which those differences can be resolved. And if we cannot end now our differences, at least we can help make the world safe for diversity. For, in the final analysis our most basic common link is that we all inhabit this small planet. We all breathe the same air. We all cherish our children's future. And we are all mortal.

Third: Let us re-examine our attitude towards the cold war, remembering we are not engaged in a debate, seeking to pile up debating points. We are not here distributing blame or pointing the finger of judgment. We must deal with the world as it is, and not as it might have been had the history of the last eighteen years been different [the period since the end of the Second World War].

We must, therefore, persevere in the search for peace in the hope that constructive changes within the Communist bloc might bring within reach solutions which now seem beyond us. We must conduct our affairs in such a way that it becomes in the Communists' interest to agree on a genuine peace. And above all, while defending our own vital interests, nuclear powers must avert those confrontations which bring an adversary to a choice of either a humiliating retreat or a nuclear war. To adopt that kind of course in the nuclear age would be evidence only of the bankruptcy of our policy—or of a collective death-wish for the world.

To secure these ends, America's weapons are non-provocative, carefully controlled, designed to deter and capable of selective use. Our military forces are committed to peace and disciplined in self-restraint. Our diplomats are instructed to avoid unnecessary irritants and purely rhetorical hostility.

For we can seek a relaxation of tensions without relaxing our guard. And, for our part, we do not need to use threats to prove that we are resolute. We do not need to jam foreign broadcasts out of fear our faith will be eroded. We are unwilling to impose our system on any unwilling—but we are willing and able to engage in peaceful competition with any people.

Meanwhile, we seek to strengthen the United Nations, to help solve its financial problems, to make it a more effective instrument for peace, to develop it into a genuine world security system—a system capable of resolving disputes on the basis of law, of insuring the security of the large and the small, and of creating conditions under which arms can finally be abolished.

At the same time we seek to keep peace inside the non-Communist world, where many nations, all of them our friends, are divided over issues which weaken Western unity, which invite Communist intervention or which threaten to erupt into war. . . .

Speaking of other nations, I wish to make one point clear. We are bound to many nations by alliances. These alliances exist because our concern and theirs substantially overlap. Our commitment to defend Western Europe and West Berlin, for example, stands undiminished because of the identity of our vital interests. The United States will make no deal with the Soviet Union at the expense of other nations and other peoples, not merely because they are our partners, but also because their interests and ours converge.

Our interests converge, however, not only defending the frontiers of freedom, but in pursuing the paths of peace.

It is our hope—and the purpose of allied policies—to convince the Soviet Union that she, too, should let each nation choose its own future, so long as the choice does not interfere with the choices of others. The Communist drive to impose their political and economic system on others is the primary cause of world tension today. For there can be no doubt that, if all nations could refrain from interfering in the self-determination of others, the peace would be much more assured.

This will require a new effort to achieve world law—a new context for world discussions. It will require increased understanding between the Soviets and ourselves. And increased understanding will require increased contact and communication.

One step in this direction is the proposed arrangement for a direct line between Moscow and Washington, to avoid on each side the dangerous delays, misunderstanding, and misreadings of the other's actions which might occur in a time of crisis.

We have also been talking in Geneva about other first-step measures of arms control, designed to limit the intensity of the arms race and reduce the risks of accidental war.

Our primary long-range interest in Geneva, however, is general and complete disarmament—designed to take place by stages, permitting parallel political developments to build the new institutions of peace which would take the place of arms. The pursuit of disarmament has been an effort of this Government since the 1920s. It has been urgently sought by the past three Administrations. And however dim the prospects are today, we intend to continue this effort—to continue it in order that all countries, including our own, can better grasp what the problems and the possibilities of disarmament are.

The only major area of these negotiations where the end is in sight—yet where a fresh start is badly needed—is in a treaty to outlaw nuclear tests. The conclusion of such a treaty—so near and yet so far—would check the spiralling arms race in one of its most dangerous areas. It would place the nuclear powers in a position to deal more effectively with one of the greatest hazards which man faces in 1963—the further spread of nuclear weapons. It would increase our security—it would decrease the prospects of war.

Surely this goal is sufficiently important to require our steady pursuit, yielding neither to the temptation to give up the whole effort nor to the temptation to give up our insistence on vital and responsible safeguards.

I am taking this opportunity, therefore, to announce two important decisions in this regard:

First: Chairman Khrushchev, Prime Minister Macmillan and I have agreed that high-level discussions will shortly begin in Moscow towards early agreement on a comprehensive test ban treaty. Our hopes must be tempered with the caution of history—but with our hopes go the hopes of all mankind.

Second: To make clear our good faith and solemn convictions on the matter, I now declare that the United States does not propose to conduct nuclear tests in the atmosphere so long as other states do not do so. We will not be the first to resume. Such a declaration is no substitute for a formal binding treaty—but I hope it will help us achieve one. Nor would such a treaty be a substitute for disarmament—but I hope it will help us achieve it.

Finally, my fellow Americans, let us examine our attitude towards peace and freedom here at home. The quality and spirit of our own society must justify and support our efforts abroad. We must show it in the dedication of our own lives—as many of you who are graduating today will have an opportunity to do, by serving without pay in the Peace Corps abroad or in the proposed National Service Corps here at home.

But wherever we are, we must all, in our daily lives, live up to the age-old faith that peace and freedom walk together. In too many of our cities today, the peace is not secure because freedom is incomplete.

It is the responsibility of the executive branch at all levels of government—local, state and national—to provide and protect that freedom for all of our citizens by all means within our authority. It is the responsibility of the legislative branch at all levels, wherever the authority is now inadequate, to make it adequate. And it is the responsibility of this country to respect the rights of others and respect the law of the land.

All this is not unrelated to world peace. "When a man's ways please the Lord," the scriptures tell us, "he maketh even his enemies to be at peace with him." And is not peace, in the last analysis, basically a matter of human rights—the right to live out our lives without fear of devastation—the right to breathe air as nature provided it—the right of future generations to a healthy existence?

While we proceed to safeguard our national interests, let us also safeguard human interests. And the elimination of war and arms is clearly in the interest of both.

No treaty, however much it may be to the advantage of all, however tightly it may be worded, can provide absolute security against the risks of deception and evasion. But it can—if it is sufficiently effective in its enforcement and it is sufficiently in the interests of its signers—offer far more security and far fewer risks than an unabated, uncontrolled, unpredictable arms race.

The United States, as the world knows, will never start a war. We do not want a war. We do not now expect a war. This generation of Americans has already had enough—more than enough—of war and hate and oppression. We shall be prepared if others wish it. We shall be alert to try to stop it. But we shall also do our part to build a world of peace where the weak are safe and the strong are just.

We are not helpless before that task or hopeless of its success. Confident and unafraid, we labor on—not towards a strategy of annihilation but towards a strategy of peace.

If Osgood's paper describes a theory of how to reduce tensions, and Kennedy's speech manifests it, Etzioni's case study examines what happened. A sociologist at Columbia University when he wrote the following chapter, Etzioni is richly experienced in public policy analysis. While not neglecting limits to "graduated reciprocation," he describes advantages of such scope that it's a wonder so little use has been made of this approach to reducing tension.

The Kennedy Experiment

Amitai Etzioni

THE KENNEDY EXPERIMENT CAN BE VIEWED AS A TEST OF a moderate version of the psychological theory that seeks to use symbolic gestures as unilateral initiatives to reduce tension to get at other factors, leading toward multilateral negotiations.

The first step was a speech by President John F. Kennedy at the American University on June 10, 1963, in which he outlined "A Strategy of Peace." While it is not known to what degree the President or his advisors were moved by a psychological theory, the speech clearly met a condition of this theory—it set the *context* for the unilateral initiatives to follow. As any concrete measure can be interpreted in a variety of ways, it is necessary to spell out the general state of mind these steps attempt to communicate.

The President called attention to the dangers of nuclear war and took a reconciliatory tone toward the Soviet Union in his address. He said that "constructive changes" in the Soviet Union "might bring within reach solutions which now seem beyond us." He stated that "our problems are man-made . . . and can be solved by man." Coming eight months after the 1962 Cuban crisis, when the United States and Russia stood "eyeball to eyeball," such statements marked a decisive change in American attitudes. United States policies, the President added, must be so constructed "that it becomes in the Communist interest to agree to a genuine peace," which was a long way from the prevailing sentiment that there was little the United States could do, so long as the Soviet Union did not change. . . . Nor did the President imply that all the blame for the cold war rested with the other side; he called on Americans to "re-examine" their attitudes toward the cold war.

Beyond merely delivering a speech, the President announced the first unilateral initiative—the United States was stopping all nuclear tests in the atmosphere and would not resume them unless another country did. This, it should be noted, was basically a psychological gesture and not a unilateral arms limitation step. The United States at that time was believed to command about five times the means of delivery of the Soviet Union and to have them much better protected, and had conducted about twice as many nuclear tests, including a recent large round of testing. American experts believed that it would take about one to two years before the information from these tests was finally digested, that in all likelihood little was to be gained from additional testing even after that date, and that if testing proved to be necessary it could be conducted in other environments, particularly underground. Thus, in effect, the President used the termination of testing as a psychological gesture.

The steps that followed had much the same quality. Kennedy's speech, delivered on June 10, was published in full during the next few days in the Soviet government newspaper, *Izvestia*, as well as in *Pravda* with a combined circulation of 10,000,000, a degree of attention rarely accorded a Western leader. Radio jammers in Moscow were turned off to allow the Russian people to listen without interruption to the Voice of America's recording of the speech, a fact that was reported in the United States and, therefore, had some tension reduction effect on both sides. Premier Khrushchev followed on June 15 with a speech welcoming the Kennedy initiative. He stated that a world war was not inevitable and that the main danger of conflict stemmed from the arms race and the stockpiling of nuclear weapons. Khrushchev reciprocated on the psychological-military side by announcing he had ordered that the production of strategic bombers be halted. The psychological nature of this step is to be seen in that the bombers were probably about to be phased out anyway and that no verification was offered for cessation of production.

In the United Nations, the Soviet Union had on June 11 removed its objection to a Western-backed proposal to send observers to war-torn Yemen. The United States had reciprocated by removing, for the first time since 1956, its objection to the restoration of full status of the Hungarian delegation to the United Nations.

Although the United States had proposed a direct America-Russia communications link at Geneva in late 1962, the Soviets finally agreed to this measure on June 20, 1963. Next, attention focused on the test ban. Following the United States' example, Russia reciprocated by not testing in the atmosphere, so that until the treaty was signed, both sides refrained from such testing under an understanding achieved without negotiation but rather through unilateral-reciprocal moves. This development, in line with the moderate version of the theory, led in July to multilateral negotiations and a treaty, signed on August 5, 1963. The signing of the treaty was followed by a number of new proposals for East-West agreements. Foreign Minister Gromyko, on September 19, 1963, called for a "non-aggression pact between the members of the Warsaw Treaty [*sic*] and the members of the North Atlantic bloc" and asked for a peace treaty with Germany. President Kennedy came before the United Nations and dramatically suggested, on September 20, 1963, that the United States and the Soviet Union explore the stars together. Also mentioned repeatedly in the front-page news in those weeks were the possible exchange of observer posts at key points to reduce the danger of surprise attack; expansion of the test treaty to include underground testing; direct flights between Moscow and New York; and the opening of an American consulate in Leningrad and a Soviet one in Chicago.

The next step actually taken came in a different area—a symbolic reduction of the trade barriers between East and West. As part of the cold war, the United States and, following its guidance, other Western nations had sharply limited the trade between East and West. Not only was trading of a long list of strategic materials forbidden, but trade in other materials required an export license that was difficult to obtain. Restrictions were also imposed on the credits Russia could obtain. There were occasional violations of these bans, especially by traders in Western countries other than the United States, but the total East-West trade remained very small.

On October 9, 1963, President Kennedy approved the sale of $250 million's worth of wheat to the Soviet Union. The almost purely psychological nature of this step is not always understood. As the test ban treaty had, for reasons mentioned above, a limited military significance, so the wheat deal had little commercial importance. The barriers to East-West trade were *not* removed; credit and license barriers were maintained. The President himself said that this decision did not initiate "a new Soviet-American trade policy," and such trade remained a small fraction of the total Soviet foreign trade. The total value of the wheat the United States actually sold was $65 million. The main values of the deal were, hence, as a gesture and in the educational effect of the public debate which preceded the Administration's approval of the deal.

October brought another transformation of a unilateral-reciprocal understanding into a binding, multilateral formal agreement. This time it concerned the orbiting of weapons of mass destruction and, once more, though it appeared to be a military measure, it was largely a psychological one. The United States had formerly decided, after considerable debate, that it was not interested in orbiting nuclear bombs. The Soviet Union, as far as could be determined, had reached a similar conclusion. Neither side orbited such weapons while it was watching the other side. On September 19 Gromyko suggested such a pact, and Kennedy indicated that the United States was willing. An agreement in principle was announced on October 3, and the final resolution was passed in the General Assembly on October 19, with the approval of both powers. Its immediate effect was to publicize and formalize an area of agreement that had in effect existed in the preceding years. Another measure, psychological in nature, was an exchange of released spies. While spies had been exchanged under a variety of circumstances in the past, the October 1963 exchange served the new policy.

In late October and in the first three weeks of November, there was a marked slow-down of American initiatives, and reciprocation to Soviet initiatives almost completely stopped. The reasons were many:

the Administration felt that the psychological mood in the West was getting out of hand, with hopes and expectations for more Soviet-American measures running too high; allies, especially West Germany, objected more and more bitterly; and the pre-election year began, in which the Administration seemed not to desire additional accommodations. The present posture seemed best for domestic purposes. There had been some promising signs for those who favored disarmament, and no matters of grave enough importance were involved so that even if all went sour—if the Soviets resumed testing, orbited bombs, etc.—no credible "appeasement" charge could be made by Republicans. There was an expectation that moves would be renewed after the [1964] elections. For the election year, however, even such measures as air and consular treaties were delayed. (The experiment was actually resumed after the election; the factors that prevented its success merit a study in their own right.)

Soviet Responses

One of the prevalent criticisms against the unilateral initiatives theory is that the Soviets might not respond to such initiatives. The Soviets, it is said, are Marxists and quite aware of the difference between real and symbolic moves. A policy of symbolic gestures would appeal only to people who think in Madison Avenue terms and not in political, military, and economic ones. The evidence on this point is fairly clear. For each move that was made, the Soviets reciprocated. Kennedy's "Strategy for Peace" speech was matched by a conciliatory speech by Khrushchev; Kennedy's unilateral declaration of cessation of tests was followed by a cessation of the production of strategic bombers; spies were traded for spies, etc. The Russians showed no difficulties in understanding the gestures and in responding to psychological initiatives; and they participated in a "you move-I move" sequence rather than waiting for simultaneous, negotiated, agree-upon moves. Further, they shifted to multilateral-simultaneous arrangements once the appropriate mood was generated, as reflected in the test-ban treaty and outer space resolution.

Another "danger" critics of unilateral initiatives warned of was that the Soviets might reciprocate "below par" and thus accumulate an advantage. While these matters are not readily measurable, it seems that the Russian reciprocations were "proportional" to the American ones. Khrushchev's speech might have been somewhat less elegant than Kennedy's, but it would be difficult to defend the proposition that announcing a halt to the production of bombers is lower in value than the declaration of cessation of tests, both basically psychological gestures. . . . The test treaty and the space ban involved substantively identical, strategically similar, commitments. In short, neither side seemed to have made a disproportionate gain.

While the warnings of the critics were not realized, a danger that seems not to have been anticipated by the United States Government did materialize: the Russians responded not just by reciprocating American initiatives but by offering some initiatives of their own, in the spirit of the *détente*. Washington was put on the spot: it had to reciprocate if it were not to weaken the new spirit, but it could lose control of the experiment. The first test came at the very outset, when Russia took the initiative and suddenly removed its objection to the sending of United Nations observers to Yemen. The United States reciprocated, as previously mentioned, allowing the restoration of full status to the Hungarian delegation to the United Nations. The United States also responded handsomely to Russia's initiative on a space ban. It found it more difficult, however, to respond to the other Russian initiatives. The United States agreed to the wheat deal, but only after hesitation that was sufficient to reduce the gesture's value. It never quite succeeded in making a good case for its objection to a non-aggression pact between the North Atlantic Treaty Organization and the Warsaw Treaty Organization. (The argument that this would involve a recognition of East Germany was a thin one, for several wordings were suggested that would circumvent this difficulty.) It was felt that a non-aggression pact between these two was already covered within the United Nations charter, which would be weakened if the message were rearticulated in another document. In other cases [however] the United States was unconcerned about such duplication, for instance, between the Organization of American States and the United Nations. The United States hesitated in responding to the Soviet initiative on an air treaty, as well as on more encompassing moves regarding Germany and disarmament. Despite this reluctance, however, there were enough initiatives and reciprocations as well as multilateral measures within the three months to allow a partial testing of the theory. What was the effect of the gestures and counter-gestures?

The Psychological Impact

The first steps in June 1963 did not produce what later became known as the Soviet-American *détente,* or the 1963-64 thaw in the cold war. In accord with the preceding psychological analysis, they were rather received with much ambivalence and suspicion. The *New York Times* seems to have reflected accurately the mood the author observed in Washington at the time, when it stated on June 16, 1963, that "there was a new threat of international peace in the air this week, the kind of threat that leaves sophisticates smirking and the rest of us just dumbfounded. The 'accommodators,' as outraged Republicans call them, were simply delighted. The 'cold warriors,' as the accommodators call them, regarded conciliation as a shrewd new tactic." Thus, even the initiating side was not convinced that there really was a new line, and, if we may assume that Russian authorities read the *New York Times*, they too could hardly have been immediately persuaded.

In line with the theory, Kennedy's initiation speech included recognition of Russia's achievements ("We can still hail the Russian people for their many achievements—in science and space, in economic and industrial growth, in culture and in acts of courage") and suffering ("And no nation in the history of battle ever suffered more than the Soviet Union suffered in the course of the Second World War"). These statements seemed to have weakened the rigid image that was typical of the cold war period.

The impact of the speech was felt outside the seats of government. In the United States, "from around the country came a generous flow of messages echoing all these responses, but more approving than not. And from around the globe came new bursts of hope kept alive by quick signs of interest in Moscow" (*NYT*, 6/16/63). A correspondent in Moscow reported that "the ready approval of its contents by ordinary Russians was evident in the reactions of Muscovites who lined up at kiosks to buy newspapers" (*NYT*, 6/16/63). But the main turning point came when the test treaty—considered an important "breakthrough"—was successfully negotiated. That at first hopes for a treaty ran low, and that it took great effort to obtain it, only increased the significance of its ratification.

The treaty to partially ban thermonuclear tests was the central gesture of the Kennedy experiment. Until it was reached, gestures and counter-gestures were met with caution, if not skepticism. When in early July Khrushchev offered a ban on tests in sea, air, and space (as was ultimately agreed), but coupled this offer with a suggestion of a non-aggression pact between the North Atlantic Treaty Organization and the Warsaw Treaty Organization, the *New York Times* referred to the offer as "Another Booby Trap?" (*NYT*, 7/7/63).

A week later, discussing the test treaty negotiations, the same source reflected the mood in the capital: "If these talks are successful, it is generally believed that a new chapter in East-West relations will open. But there are grave doubts on all sides that such a new chapter is indeed at hand" (*NYT*, 7/14/63). Thus, a test ban was viewed as having major tension-reduction potential, but there was much doubt whether it would be achieved. A Washington reporter still refers to the *détente* at this point with a question mark and explores at length the possibility "that the Soviet Union did not really want an agreement," that it was negotiating in bad faith (*NYT*, 7/14/63). An American report from Moscow indicated that "Mr. Khrushchev would also hope that conclusion of a partial test-ban treaty would create an atmosphere in which he could negotiate other advantageous agreements, expecially on Germany."

The treaty was negotiated in July, signed in August, and ratified in September. Thus, for more than two months, it served as the focus for discussions about Soviet intentions, the possibility of peaceful coexistence, and the dangers of nuclear war; and the Senate hearing helped to keep the debate alive. Its ratification was therefore not merely one more gesture in an international sequence of pseudo-events, but a major educational act. The American public that entered the period with ambivalent attitudes toward a test-ban treaty, remembering the arbitrary resumption of testing by the Soviet Union in 1961, after three years of voluntary moratorium, as well as the 1962 Cuban crisis, was now strongly in favor of the agreement. Louis Harris reports that a national poll taken in July, before the negotiations on the treaty had begun, found that 52 per cent of the population strongly supported a treaty. This percentage had risen to 81 by September when the treaty was ratified (*Washington Post*, 9/16/63). The tone of the press also changed; there was now an "official amity" between the United States and the Soviet Union (*NYT*, 8/4/63). While some newspapermen, accustomed to sudden shifts in international winds, continued to be cautious, a report from Moscow stated: "As Secretary of State Rusk left the Soviet Union today, after six days of discussions with Soviet leaders, it appeared almost certain to Western observers here that a surface of calm would descend on East-West relations. . . . The prospect, it is believed, is for a long period of manifold negotiations at all levels and in many cities and countries on all sorts of issues. . . . The feeling is that the Russians are generally interested in maintaining the current state of improved relations with the West. They are believed to be hoping for a minimum of friction" (*NYT*, 8/11/63). The correspondent who had reported smirking and dumbfoundedness over any possible thaw in June now stated that "we have cleared the air and cleared the atmospheres and warmed the climate and calmed the winds." The test-ban treaty had allayed many of the doubts about Russian intentions.

Following the signing of the treaty came a number of new proposals to improve East-West relations and further extend the *détente*. While none of these materialized in this period, the repeated and frequent offering of various tension-reduction measures had some effect in itself. Actually, hopes rose so quickly that late in August, Secretary of Defense McNamara warned that it was perilous to relax in a "euphoria," and Kennedy cautioned in September that the test ban was "not the millennium."

By late October, almost no new American initiatives were taken, and those of the Soviet Union were not reciprocated. The press referred to a "pause in the thaw"; there was a marked slow-down in tension reduction though efforts continued to preserve the measure of *détente* that had been achieved. The assassination of President Kennedy and the beginning of the election year ushered in a year of more or less stable semi-*détente.*

What are the conclusions from this brief and incomplete test of the theory? Certain of the central hypotheses were supported: (a) unilateral gestures were reciprocated; (b) reciprocations were proportional; (c) unilaterally reciprocated gestures reduced tensions; (d) unilaterally reciprocated gestures were followed by multilateral-simultaneous measures, which further reduced tensions; (e) initiatives were "suspected," but, when continued, they "got across"; (f) the gestures and responses created a psychological momentum that pressured for more measures, a reversal of the cold war or hostility spiral; (g) when measures were stopped, tension reduction ceased; (h) the relatively more consequential acts were initiated multilaterally or were transformed from an initially informal, unilaterally reciprocated basis to a formal, mutilateral one.

Not all the assumptions and derivations of the theory were as clearly supported. Most important, it is impossible to tell, without rerunning history for "control" purposes, whether multilateral negotiations could have been successfully undertaken without the "atmosphere" first having been improved by unilateral steps. The fact, however, that both the test-ban treaty and the space ban were first introduced on a unilateral-reciprocal basis and that even in the reduced tension condition these measures were hard to defend before Congress suggests that, if not preceded by tension reduction, they either might have failed, or the risks of failure would have been sufficiently high for the Administration to refrain from introducing them. (Attempts to advance a test ban in earlier periods failed.)

Also, the Kennedy experiment was only a partial application of the theory: the gestures were not the clear signals a full test of theory would require. Thus, for instance, to gain the Senate's consent for a test-ban treaty, its value for American security was stressed. It would allow, it was said, stopping of testing while we were ahead both in number of tests and weapons technology. Further, President Kennedy made it clear that the United States would "vigorously and diligently" pursue its underground nuclear test program. The wheat deal was interpreted in a similar fashion, e.g., as a show of Russia's weakness. Further, during the whole period, American observers provided various interpretations of the gestures as other than efforts to communicate a desire for peaceful co-existence (e.g., the *détente* exacerbates the Soviet-Sino rift). While a policy is often supported by a large variety of arguments, and the self-serving ones are usually emphasized when facing Congress, their preponderance could not but have had negative side-effects on Soviet-American relations. Also, the same gestures would have been more effective had they been introduced with less hesitation, and if Soviet initiatives had been met with less ambivalence.

To what purpose could a policy of "graduated reciprocation" be applied today? Some say it could help in creating technical and political arrangements that would lessen the chances of nuclear weapons being *used.* Others, such as businessman Harold Willens, want to deal directly with the hardware and—in stages—stop such weapons from being tested, deployed, or produced. Then the stockpiles could be reduced. Toward this end, Willens, a leader in the "nuclear freeze" movement, urges a strategy of "independent initiatives," citing Kennedy's American University speech.

Of course, there's nothing magic about initiatives. Gorbachev announced (and later extended) a moratorium on Soviet nuclear tests without inducing the U.S. to halt tests that the Reagan Administration says are vital to its "Star Wars" research program. Taking an initiative that you *know* the other side will reject constitutes propaganda instead of the first step toward an agreement. The trick is to identify an initiative which *both* sides could recognize as beneficial and which thus might elicit a positive response. The steps outlined by Willens are one approach. Many others also deserve to be considered.

The Next Initiatives

Harold Willens

NOW IS THE TIME FOR THE UNITED STATES TO TAKE *independent initiatives* that will break the arms control deadlock and accelerate the arms reduction process. As the leader of the free world and the leader in the nuclear weapons race, we should take action that is in our self-interest and made possible by our democratic system of government. We can demonstrate to the world our resourcefulness and imagination, our willingness to experiment and take a carefully calculated risk that is minimal when compared to the enormous risk of nuclear war.

There is a powerful precedent for this kind of leadership and American independent action. Twenty years ago President John F. Kennedy, speaking at the American University commencement, made the bold declaration that the United States would not conduct any further atmospheric nuclear testing as long as other nations would also refrain from testing. . . .

Setting Positive Mutual Examples

Once again it is time for the United States to exercise leadership and take bold action. **We should begin immediately a step-by-step program of American initiatives aimed at slowing, stopping, and reversing the Soviet-American nuclear arms race.** This incremental program—a building-block approach to halting the arms race—would include the following steps:

Step One

The United States would announce that we will suspend any further testing of nuclear weapons and challenge the Soviet Union to reciprocate while we conclude the negotiation of a formal Comprehensive Test Ban Treaty. If the Soviets continued testing or refused to conclude the Treaty within a reasonable period of time, the U.S. would no longer be bound to its proposal.

A Comprehensive Test Ban Treaty has been in negotiation since 1963, but the nuclear weapons establishment—a loose network of scientists, military men, nuclear war planners, and some members of Congress—has lobbied actively and successfully to block any agreement to suspend testing. President Reagan formally abandoned all efforts to negotiate the treaty in 1982.

Over the past 24 years, the issue of verification has been a major stumbling block in concluding a treaty. Although it is much easier to detect atmospheric testing, there have been significant advances in the ability to monitor underground tests as well. Independent scientific researchers have reported that "the technical capabilities needed to police a comprehensive test ban down to explosions of very small size unquestionably exist." There is the risk, of course, that small violations might go undetected, but these small risks must be weighed against the very positive advantage of taking a mutual step that could lead us out of the Soviet-American nuclear weapons impasse.

A moratorium on nuclear testing would be a realistic and quickly achievable first step in U.S. independent initiatives. A positive response by the Soviet Union would create a climate in which we could finally reach agreement on the long-sought Comprehensive Test Ban Treaty. Former Director of the Arms Control and Disarmament Agency Paul Warnke stated in June of 1983 that negotiations on the Comprehensive Test Ban were so close to completion that, if the United States and the Soviet Union were both willing, the Treaty could be signed within a period of thirty days.

If an American President were unwilling to propose such a moratorium on weapons testing, the Congress could take the initiative by amending the relevant appropriations bill or by other legislative action. Provision would be made, of course, for resumption of U.S. testing if Soviet tests continued. This first step is a no-risk, high-yield approach that—contingent upon Soviet response—could produce a significant breakthrough. We have nothing to lose and everything to gain.

Step Two

After accomplishing the critical first step of a test-ban agreement, we would have established some degree of mutual confidence along with procedures that would make successive steps less risky and less difficult. Our next independent initiative could be to propose a moratorium on the flight-testing of nuclear weapons delivery systems. Like the ban on testing, this step would be relatively easy to verify, since flight-testing of missiles and aircraft is highly visible.

Such a moratorium, again contingent upon Soviet compliance, would prevent the testing and thus inhibit the deployment of highly destabilizing delivery systems like the new family of ICBMs—our MX missile and whatever the Soviets will come up with as its counterpart. It would also inhibit deployment of the cruise missile as well as a whole new generation of very accurate sea-launched weapons.

As with each step of this proposal, a successful moratorium would be codified into formal agreements that would build confidence on both sides. Each would gradually strengthen its belief that the other was not adding to its nuclear capabilities.

Step Three

Building on the momentum of this sequential, incremental process, the United States should propose yet another moratorium on the deployment of any new nuclear weapons systems. As with testing, verification of this step is not difficult, since even now we are able to count every Soviet missile launcher, bomber, and submarine. Again, we would announce our own halt in deployment, challenging the Russians to do the same. In negotiating a formal agreement, we should also seek to make certain that both sides maintain their commitment to already agreed SALT I and II limits on the deployment of existing weapons systems.

Step Four

With the first three steps successfully taken, we could then begin to face the far more difficult issue of banning the production of fissionable material and the actual production of nuclear warheads. Stopping production is more difficult to verify, since the manufacturing process is easier to conceal. But with the first three steps accomplished, there would be little motivation to continue making the weapons. It is also likely that there will be further advances in verification procedures. We should keep in mind that the Soviets have already formally accepted the principle of on-site inspection for arms control verification as part of a comprehensive test ban. The first three steps certainly would establish still more mutual confidence, a greater degree of trust, and implementation of improved verification procedures by both sides.

Stopping the production of new nuclear weapons offers several advantages to both sides: 1) It would stop the production of an entire generation of new weapons, now in development, which are dangerous and destabilizing; 2) It would prevent both sides from deploying additional cruise missiles that, because they are small and easily concealed, would make future arms control verification difficult, if not impossible; 3) It would keep the arms race from moving into outer space; and 4) Each side would still retain an effective nuclear deterrent capacity.

Impartial experts (among them, William Colby, former Director of the Central Intelligence Agency, Jeremy Stone, Director of the American Federation of Scientists, and former CIA Deputy Director Herbert Scoville, now president of the Arms Control Association) agree that a halt in production is at least as verifiable as other arms agreements, if not more so, since it is much more difficult to verify complex numerical ceilings on various types of weapons than it is to detect evidence of significant weapons production. In testimony before Congress, former CIA Director Colby stated that "any program which offered the prospect of strategic advantage to the Soviets by definition would have to be of a size and consequent visibility that we could identify it long before it became a direct threat, and take defensive action against it."

In essence, Steps One, Two, Three, and Four of this proposal amount to an incremental nuclear weapons freeze—a step-by-step process for accomplishing the aims now set forth in the 1982 Bilateral Nuclear Weapons Freeze Initiative—a simple, straightforward proposal calling for an immediate halt to testing, production, and deployment of nuclear weapons by the United States and the Soviet Union. The freeze was supported by more than twelve million voters in nine states. It has been described as the largest referendum on a single issue in the history of the United States, and citizen support for its common-sense clarity continues to grow. Six months after the 1982 election results, a national poll showed that 79 percent of Americans favored the freeze. Its bipartisan support was demonstrated by the fact that 72 percent of the Republicans polled favored the freeze. Even the noted conservative columnist James Kilpatrick has stated that "there is nothing in the freeze resolution that a good conservative could not support."

The step-by-step approach proposed here, beginning with the moratorium on testing, could accomplish the aims of the widely favored Sovet-American freeze without necessitating lengthy and tedious negotiation of every aspect all in one lump. The most pressing need is to begin immediately the process of slowing, stopping, and reversing a nuclear arms race that continues to escalate.

Step Five

Once the United States and the Soviet Union agree to halt future weapons development and to ensure that neither side could get ahead in weaponry, the ground would be laid for true reductions in the massive nuclear arsenals of both nations. Step Five in the process proposed here would begin with an appraisal of existing arsenals and the proposal of positive measures to carry out a stable, orderly, and balanced program of reductions. Each side would keep an adequate deterrent capability while lowering the level of its nuclear arsenal.

Former Ambassador to the Soviet Union George Kennan has made a proposal for massive reductions that is sensible, simple, and gets right to the heart of the matter. Kennan proposes an immediate, across-the-board 50-percent reduction in the nuclear arsenals of both superpowers, "to be implemented at once without further wrangling among the experts." The 50 percent left on each side would still be far more than enough to devastate the adversary.

Admiral Noel Gayler, former Director of the National Security Agency, has expanded upon Ambassador Kennan's proposal, suggesting practical ways it might be carried out. He believes that deep cuts of this nature are entirely feasible, that each side could simply turn in (to a joint U.S.-Soviet commission or an impartial third party) an equal number of nuclear devices. "Let each side choose the weapons it wishes to turn in, whether missile warheads, bombs or artillery shells. Each weapon would count the same—as one device. . . . A nuclear device is uniquely identifiable and can be counted without error when turned in; thus, there is full verification without intrusive inspection in either country."

Admiral Gayler's suggestion is to have the process begin with a small number of weapons—say, fifty—to test the method and develop confidence, and then continue on an agreed schedule toward very large reductions of perhaps ten thousand devices from each side. Both countries would still retain a small number of strategic weapons in reserve. Again, the risks involved are far outweighed by the immense risk of continuing to increase the destructive capability of arsenals that are already obscenely excessive.

This step-by-step program of American initiatives—the proposal of a moratorium first on weapons testing, then on testing of delivery systems, then on deployment and production, and finally the proposal of massive cuts in nuclear arsenals—represents a methodical process for breaking the momentum of the nuclear arms race with no real risk to our national security. In each step, our country could take independent action that would demonstrate good faith and give the Soviets an opportunity to demonstrate theirs as well. Independent initiatives by the United States are a pragmatic, incremental approach to a problem that too many of us have seen as having no solution. Such incrementalism is the essence of sound business practice, with each stage forming a solid basis for the following one. Step by step, our country can lead the way out.

If the United States were to lead the way, would the Soviet Union follow us? There is growing evidence that Soviet leadership desires serious negotiations to reduce the growing threat of annihilation and the crushing burden of the arms race. The Soviet economy, far less efficient and productive than that of the United States, has a host of economic problems—including failing agriculture, a shrinking work force, shortages of consumer goods, and backward technology—that plague the Soviet system. Increasing trouble with Russia's ethnic minorities, unrest in Poland, and a continuing war in Afghanistan add to the problems the Soviet leadership cannot ignore. Reducing the drain of the arms race on their economy would enable the USSR to focus on serious domestic problems and perhaps curb its appetite for international adventurism.

As is the case with any competitor, it is unrealistic to expect the Soviets to act in our interest. But it is perfectly realistic to expect them to act in their own self-interest, if given the opportunity. Reversing the nuclear arms race is as much in their interest as it is in ours.

Even if a change is begun through independent initiatives, it will, in most cases, eventually require negotiations. (In fact, a cycle of initiatives can be viewed as a form of negotiating.) In their best-selling book, *Getting to Yes*, Roger Fisher and William Ury stress a distinction between "positional bargaining" and "negotiation on the merits." The first is the method of the rug bazaar or second-hand car lot. The second approach keeps asking, "What are the real underlying interests here?" It tries to expand the list of possibilities instead of settling for a compromise between opening positions. It proposes objective criteria for settling disputes.

An excellent illustration of the used-car approach is provided by a dispute in the history of arms control. For years, the U.S. and the Soviet Union debated whether to allow "on-site inspections" to verify compliance. At one point, the U.S. wanted twenty of these on-the-ground inspections, or at least a dozen, while the Soviets, who wanted none at all, seemed open to allowing a few. It is not at all clear, however, that anybody knew what an on-site inspection would include. Instead, the sides were haggling over numbers. It's as if you went to a used car dealer and bargained over the price before picking a car. Fisher and Ury tell not only how to avoid such absurdities and reach an agreement, but also how to use their method to strengthen your working relationship with the other side.

Getting to Yes

Roger Fisher
William Ury

WHETHER A NEGOTIATION CONCERNS A CONTRACT, A family quarrel, or a peace settlement among nations, people routinely engage in positional bargaining. Each side takes a position, argues for it, and makes concessions to reach a compromise. The classic example of this negotiating minuet is the haggling that takes place between a customer and the proprietor of a secondhand store.

Customer: How much do you want for this brass dish?

Shopkeeper: That is a beautiful antique, isn't it? I guess I could let it go for $75.

C: Oh come on, it's dented. I'll give you $15.

S: Really! I might consider a serious offer, but $15 certainly isn't serious.

C: Well, I could go to $20, but I would never pay anything like $75. Quote me a realistic price.

S: You drive a hard bargain, young lady. $60 cash, right now.

C: $25.

S: It cost me a great deal more than that. Make me a *serious* offer.

C: $37.50. That's the highest I will go.

S: Have you noticed the engraving on that dish? Next year pieces like that will be worth twice what you pay today.

And so it goes, on and on. Perhaps they will reach agreement; perhaps not.

Any method of negotiation may be fairly judged by three criteria: It should produce a wise agreement if agreement is possible. It should be efficient. And it should improve or at least not damage the relationship between the parties. (A wise agreement can be defined as one which meets the legitimate interests of each side to the extent possible, resolves conflicting interests fairly, is durable, and takes community interests into account.)

The most common form of negotiation, illustrated by the above example, depends upon successively taking—and then giving up—a sequence of positions.

Taking positions, as the customer and storekeeper do, serves some useful purposes in a negotiation. It tells the other side what you want; it provides an anchor in an uncertain and pressured situation; and it can eventually produce the terms of an acceptable agreement. But those purposes can be served in other ways. And positional bargaining fails to meet the basic criteria of producing a wise agreement, efficiently and amicably.

Arguing over Positions Produces Unwise Agreements

When negotiators bargain over positions, they tend to lock themselves into those positions. The more you clarify your position and defend it against attack, the more committed you become to it. The more you try to convince the other side of the impossibility of changing your opening position, the more difficult it becomes to do so. Your ego becomes identified with your position. You now have a new interest in "saving face"—in reconciling future action with past positions—making it less and less likely that any agreement will wisely reconcile the parties' original interests.

The danger that positional bargaining will impede a negotiation was well illustrated by the breakdown of the talks under President Kennedy for a comprehensive ban on nuclear testing. A critical question arose: How many on-site inspections per year should the Soviet Union and the United States be permitted to make within the other's territory to investigate suspicious seismic events? The Soviet Union finally agreed to three inspections. The United States insisted on no less than ten. And there the talks broke down—over positions—despite the fact that no one understood whether an "inspection" would involve one person looking around for one day, or a hundred people prying indiscriminately for a month. The parties had made little attempt to design an inspection procedure that would reconcile the United States's interest in verification with the desire of both countries for minimal intrusion.

As more attention is paid to positions, less attention is devoted to meeting the underlying concerns of the parties. Agreement becomes less likely. Any agreement reached may reflect a mechanical splitting of the difference between final positions rather than a solution carefully crafted to meet the legitimate interests of the parties. The result is frequently an agreement less satisfactory to each side than it could have been.

Arguing over Positions Is Inefficient

The standard method of negotiation may produce either agreement, as with the price of a brass dish, or breakdown, as with the number of on-site inspections. In either event, the process takes a lot of time.

Bargaining over positions creates incentives that stall settlement. In positional bargaining you try to improve the chance that any settlement reached is favorable to you by starting with an extreme position, by stubbornly holding to it, by deceiving the other party as to your true views, and by making small concessions only as necessary to keep the negotiation going. The same is true for the other side. Each of those factors tends to interfere with reaching a settlement promptly. The more extreme the opening positions and the smaller the concessions, the more time and effort it will take to discover whether or not agreement is possible.

The standard minuet also requires a large number of individual decisions as each negotiator decides what to offer, what to reject, and how much of a concession to make. Decision-making is difficult and time-consuming at best. Where each decision not only involves yielding to the other side but will likely produce pressure to yield further, a negotiator has little incentive to move quickly. Dragging one's feet, threatening to walk out, stonewalling, and other such tactics become commonplace. They all increase the time and costs of reaching agreement as well as the risk that no agreement will be reached at all.

Arguing over Positions Endangers an Ongoing Relationship

Positional bargaining becomes a contest of will. Each negotiator asserts what he will and won't do. The task of jointly devising an acceptable solution tends to become a battle. Each side tries through sheer will power to force the other to change its position. "I'm not going to give in. If you want to go to the movies with me, it's *The Maltese Falcon* or nothing." Anger and resentment often result as one side sees itself bending to the rigid will of the other while its own legitimate concerns go unaddressed. Positional bargaining thus strains and sometimes shatters the relationship between the parties. Commercial enterprises that have been doing business together for years may part company. Neighbors may stop speaking. Bitter feelings generated by one such encounter may last a lifetime.

When There Are Many Parties, Positional Bargaining Is Even Worse

Although it is convenient to discuss negotiation in terms of two persons, you and "the other side," in fact, almost every negotiation involves more than two persons. Several different parties may sit at the table, or each side may have constituents, higher-ups, boards of directors, or committees with whom they must deal. The more people involved in a negotiation, the more serious the drawbacks to positional bargaining.

If some 150 countries are negotiating, as in various United Nations conferences, positional bargaining is next to impossible. It may take all to say yes, but only one to say no. Reciprocal concessions are difficult: To whom do you make a concession? Yet even thousands of bilateral deals would still fall short of a multilateral agreement. In such situations, positional bargaining leads to the formation of coalitions among parties whose shared interests are often more symbolic than substantive. At the United Nations such coalitions produce negotiations between "the" North and "the" South, or between "the" East and "the" West. Because there are many members in a group, it becomes more difficult to develop a common position. What is worse, once they have painfully developed and agreed upon a position, it becomes much harder to change it. Altering a position proves equally difficult when additional participants are higher authorities who, while absent from the table, must nevertheless give their approval.

Being Nice Is No Answer

Many people recognize the high costs of hard positional bargaining, particularly on the parties and their relationship. They hope to avoid them by following a more gentle style of negotiation. Instead of seeing the other side as adversaries, they prefer to see them as friends. Rather than emphasizing a goal of victory, they emphasize the necessity of reaching agreement. In a soft negotiating game the standard moves are to make offers and concessions, to trust the other side, to be friendly, and to yield as necessary to avoid confrontation.

The soft negotiating game emphasizes the importance of building and maintaining a relationship. Within families and among friends much negotiation takes place in this way. The process tends to be efficient, as least to the extent of producing results quickly. As each party competes with the other in being more generous and more forthcoming, an agreement becomes highly likely. But it may not be a wise one. The results may not be as tragic as in the O. Henry story about an impoverished couple in which the loving wife sells her hair in order to buy a handsome chain for her husband's watch, and the unknowing husband sells his watch in order to buy beautiful combs for his wife's hair. However, any negotiation primarily concerned with the relationship runs the risk of producing a sloppy agreement.

More seriously, pursuing a soft and friendly form of positional bargaining makes you vulnerable to someone who plays a hard game of positional bargaining. In positional bargaining, a hard game dominates a soft one. If the hard bargainer insists on concessions and makes threats while the soft bargainer yields in order to avoid confrontation and insists on agreement, the negotiating game is biased in favor of the hard player. The process will produce an agreement, although it may not be a wise one. It will certainly be more favorable to the hard positional bargainer than to the soft one. If your response to sustained, hard positional bargaining is soft positional bargaining, you will probably lose your shirt.

There Is an Alternative

If you do not like the choice between hard and soft positional bargaining, you can change the game.

The game of negotiation takes place at two levels. At one level, negotiation addresses the substance; at another, it focuses—usually implicitly—on the procedure for dealing with the substance. The first negotiation may concern your salary, the terms of a lease, or a price to be paid. The second negotiation concerns how you will negotiate the substantive question: by soft positional bargaining, by hard positional bargaining, or by some other method. This second negotiation is a game about a game—a "meta-game." Each move you make within a negotiation is not only a move that deals with rent, salary, or other substantive questions; it also helps structure the rules of the game you are playing. Your move may serve to keep the negotiations within an ongoing mode, or it may constitute a game-changing move.

This second negotiation by and large escapes notice because it seems to occur without conscious decision. Only when dealing with someone from another country, particularly someone with a markedly different cultural background, are you likely to see the necessity of establishing some accepted process for the substantive negotiations. But whether consciously or not, you are negotiating procedural rules with every move you make, even if those moves appear exclusively concerned with substance.

The answer to the question of whether to use soft positional bargaining or hard is "neither." Change the game. At the Harvard Negotiation Project we have been developing an alternative to positional bargaining: a method of negotiation explicitly designed to produce wise outcomes efficiently and amicably. This method, called *principled negotiation* or *negotiation on the merits*, can be boiled down to four basic points.

These four points define a straightforward method of negotiation that can be used under almost any circumstance. Each point deals with a basic element of negotiation, and suggests what you should do about it.

People: Separate the people from the problem.
Interests: Focus on interests, not positions.
Options: Generate a variety of possibilities before deciding what to do.
Criteria: Insist that the result be based on some objective standard.

The first point responds to the fact that human beings are not computers. We are creatures of strong emotions who often have radically different perceptions and have difficulty communicating clearly. Emotions typically become entangled with the objective merits of the problem. Taking positions just makes this worse because people's egos become indentified with their positions. Hence, before working on the substantive problem, the "people problem" should be disentangled from it and dealt with separately. Figuratively if not literally, the participants should come to see themselves as working side by side, attacking the problem, not each other. Hence the first proposition: *Separate the people from the problem.*

The second point is designed to overcome the drawback of focusing on people's stated positions when the object of a negotiation is to satisfy their underlying interests. A negotiating position often obscures what you really want. Compromising between positions is not likely to produce an agreement which will effectively take care of the human needs that led people to adopt those positions. The second basic element of the method is: *Focus on interests, not positions.*

The third point responds to the difficulty of designing optimal solutions while under pressure. Trying to decide in the presence of an adversary narrows your vision. Having a lot at stake inhibits creativity. So does searching for the one right solution. You can offset these constraints by setting aside a designated time within which to think up a wide range of possible solutions that advance shared interests and creatively reconcile differing interests. Hence the third basic point: Before trying to reach agreement, *invent options for mutual gain.*

Where interests are directly opposed, a negotiator may be able to obtain a favorable result simply by being stubborn. That method tends to reward intransigence and produce arbitrary results. However, you can counter such a negotiator by insisting that this single say-so is not enough and that the agreement must reflect some fair standard independent of the naked will of either side. This does not mean insisting that the terms be based on the standard you select, but only that some fair standard such as market value, expert opinion, custom, or law determine the outcome. By discussing such criteria rather than what the parties are willing or unwilling to do, neither party need give in to the other; both can defer to a fair solution. Hence the fourth basic point: *Insist on using objective criteria.*

The method of principled negotiation is contrasted with hard and soft positional bargaining in the table below, which shows the four basic points of the method in boldface type.

The four propositions of principled negotiation are relevant from the time you begin to think about negotiating until the time either an agreement is reached or you decide to break off the effort. That period can be divided into three stages: analysis, planning, and discussion.

During the *analysis* stage you are simply trying to diagnose the situation—to gather information, organize it, and think about it. You will want to consider the people problems of partisan perceptions, hostile emotions, and unclear communication, as well as to identify your interests and those of the other side. You will want to note options already on the table and identify any criteria already suggested as a basis for agreement.

During the *planning* stage you deal with the same four elements a second time, both generating ideas and deciding what to do. How do you propose to handle the people problems? Of your interests, which are most important? And what are some realistic objectives? You will want to generate additional options and additional criteria for deciding among them.

Again during the *discussion* stage, when the parties communicate back and forth, looking toward agreement, the same four elements are the best subjects to discuss. Differences in perception, feelings of frustration and anger, and difficulties in communication can be acknowledged and addressed. Each side should come to understand the interests of the other. Both can then jointly generate options that are mutually advantageous and seek agreement on objective standards for resolving opposed interests.

To sum up, in contrast to positional bargaining, the principled negotiation method of focusing on basic interests, mutually satisfying options, and fair standards typically results in a *wise* agreement. The method permits you to reach a gradual consensus on a joint decision *efficiently* without all the transactional costs of digging in to positions only to have to dig yourself out of them. And separating the people from the problem allows you to deal directly and empathetically with the other negotiator as a human being, thus making possible an *amicable* agreement.

Problem Positional Bargaining: Which Game Should You Play?		Solution Change the Game— Negotiate on the Merits
Soft	**Hard**	**Principled**
Participants are friends.	Participants are adversaries.	Participants are problem-solvers.
The goal is agreement.	The goal is victory.	The goal is a wise outcome reached efficiently and amicably.
Make concessions to cultivate the relationship.	Demand concessions as a condition of the relationship.	**Separate the people from the problem.**
Be soft on the people and the problem.	Be hard on the problem and the people.	Be soft on the people, hard on the problem.
Trust others.	Distrust others.	Proceed independent of trust.
Change your position easily.	Dig in to your position.	**Focus on interests, not positions.**
Make offers.	Make threats.	Explore interests.
Disclose your bottom line.	Mislead as to your bottom line.	Avoid having a bottom line.
Accept one-sided losses to reach agreement.	Demand one-sided gains as the price of agreement.	**Invent options for mutual gain.**
Search for the single answer: the one *they* will accept.	Search for the single answer: the one *you* will accept.	Develop multiple options to choose from; decide later.
Insist on agreement.	Insist on your position.	Insist on using objective criteria.
Try to avoid a contest of will.	Try to win a contest of will.	Try to reach a result based on standards independent of will.
Yield to pressure.	Apply pressure.	Reason and be open to reasons; yield to principle, not pressure.

2 Developing Institutions That Favor Peace

Despite having learned how to settle disputes amicably or at least non-violently in many areas of life, humans have nonetheless continued to slaughter one another in the peculiar practice known as war. If war is so ghastly in its results, why have groups entered into it so often? Historians and participants have suggested many answers.

One is that nations, having made elaborate preparations to fight, and having become entangled in complex networks of obligation, often simply blunder into war. Another explanation, not contradictory to the first, is that training for war, belonging to an army, and the activity of fighting itself, despite danger and destruction, are in some ways deeply satisfying—a view scandalously expressed by William Broyles, Jr., in the November 1984 *Esquire.* While allowing that war is "ugly, horrible, evil, and it is reasonable to hate all that," Broyles came to believe, while reflecting upon his experience in Vietnam, that "most men who have been to war would have to admit they loved it too."

Back in 1910, just before the First World War, the psychologist and philosopher William James published an essay called "The Moral Equivalent of War." While declaring his own commitment to "a reign of peace," he asked what it is that men love about war, or at least about identifying with military values, and how some of the same values could be served without killing. Like the framers of the Republic, James did not try to suppress impulses that seemed to be causing trouble, but rather to harness them to the public good, almost regardless of individual motivation.

As in so much of his work, James here combined what he called tough-minded and tender-minded approaches. Most people who oppose war would hesitate to acknowledge that it has *any* attractions, and most of those who have felt its appeals do not believe that war *can* be avoided, even if they believe it should be. In contrast, James undertook to present, as if from the inside, the positive values

honored by militarists, and then to suggest the development of social institutions that could call upon and satisfy these values without causing destruction or giving vent to hatred.

It was a contemporary of James, Theodore Roosevelt, who praised "the strenuous life," a phrase echoed much later by John F. Kennedy when he spoke about "vigor," as he often did in his 1960 campaign. When James proposed offering the young of "the luxurious classes" an opportunity to go off "to fishing fleets in December . . . to road-building and tunnel-making . . . to get the childishness knocked out of them," we already hear one of the themes that led to the Peace Corps.

In place of the challenge, comradeship, and excitement of war, James wanted to encourage a fiery but peaceful civic passion. "It is only a question of blowing on the spark till the whole population gets incandescent," he said, "and on the ruins of the old morals of military honor, a stable system of morals of civic honor builds itself up."

Only four years after the James essay was printed, the guns of August opened fire. Many historians now regard the First World War as a classic case of statesmen blundering into a catastrophe that nobody wanted; but as we learn from some of the best memoirs, novels, and poetry of that war, it was also widely welcomed, at the start, as an escape from the tedium and soft values of civilian life.

James knew the first rule of political life, that you can't fight something with nothing—you can't reliably oppose concrete interests or deep human feelings with an abstract appeal. As for idealism, it needs to be organized. Lit in a dish, a pool of gasoline burns pointlessly; sprayed into an engine, it drives a truck up a hill. Soon after John Kennedy exhorted citizens to "ask what you can do for your country," he created the Peace Corps as an engine into which their energy could be directed.

Actually, the volunteers who began to go abroad in the early 1960s were working not so much for peace between East and West, as for local development in countries that felt exploited by colonialism and its aftermath. If the less developed nations are called the South and the industrialized ones are called the North (including such anomalies as Australia, Brazil, and South Africa), then the Peace

Corps was the North helping the South. Or rather, it was *part* of the North—the capitalist, democratic nations represented by the U.S.

Along with some other social inventers, Robert Fuller proposes a new kind of Peace Corps, one that directly addresses East-West tensions. In effect, it's a large-scale exchange program between many countries, including the U.S. and the Soviet Union. On at least one occasion President Reagan has offered a related vision.[1]

Sketchy as these proposals remain, they illustrate the need to invent institutions that favor peace. If we were to assess inventions not solely by their payoff in the consumer market, or by their technical "sweetness," but by their value to humankind, we'd quickly see that the limiting factor in the world today is the low level of social inventiveness. All the more reason to support the experimentalism and daring necessary to it, and to celebrate the examples of it that we have.

—Craig Comstock

1. See introduction to section on "Getting to Know the Other Side" in *Citizen Summitry: Keeping the Peace When It Matters Too Much to be Left to Politicians,* edited by Don Carlson and Craig Comstock (Tarcher, 1986).

During the crisis caused by high oil prices, President Carter announced that energy conservation was the moral equivalent of war. It's hard to know whether his speechwriter had a clear idea of what James meant by this highly quotable phrase. James understood that the activities of fighting, belonging to the armed forces, or threatening to fight, offer satisfactions which could be provided by other means. It is these other means that qualify as moral equivalents.

In the case of high energy prices, it seems doubtful that insulating your attic or driving at 55 m.p.h. could satisfy the longing, so widely felt at the time, to punch a sheik in the nose. What Carter may have meant is that if we conserved energy, demand would fall, and price along with it; we'd become less dependent upon imports and thus less vulnerable, as Kissinger had put it, to actual strangulation. James, however, was dealing not with material incentives to war, but with psychological motivations.

Despite several revivals of interest in his work, William James is an academic orphan. He wrote very well, but English departments prefer his brother Henry. He developed a philosophy of pragmatism, but philosophers prefer to discuss Pierce, another American original. James brilliantly described states of consciousness and religious experience, but his findings had little place in a psychology trying to become "scientific."

If James were writing today, he would surely want to take account of the intellectualism and abstraction that reduce conflict to the status of a computer simulation, thus ignoring both what Clausewitz called the "friction" of war—the messiness and unpredictability of it—and the context of human relations in which fighting is only one alternative. If he were to discuss the folly of academic strategists, however, James would surely delve into the satisfactions provided by their methods, suggesting tasks to which these methods might properly and safely be applied.

The Moral Equivalent of War

William James

THE WAR AGAINST WAR IS GOING TO BE NO HOLIDAY excursion or camping party. The military feelings are too deeply grounded to abdicate their place among our ideals until better substitutes are offered than the glory and shame that come to nations as well as to individuals from the ups and downs of politics and the vicissitudes of trade. There is something highly paradoxical in the modern man's relation to war. Ask all our millions, north and south, whether they would vote now (were such a thing possible) to have our war for the Union expunged from history, and the record of a peaceful transition to the present time substituted for that of its marches and battles, and probably hardly a handful of eccentrics would say yes. Those ancestors, those efforts, those memories and legends, are the most ideal part of what we now own together, a sacred spiritual possession worth more than all the blood poured out. Yet ask those same people whether they would be willing in cold blood to start another civil war now to gain another similar possession, and not one man or woman would vote for the proposition. In modern eyes, precious though wars may be, they must not be waged solely for the sake of the ideal harvest. Only when forced upon one, only when an enemy's injustice leaves us no alternative, is a war now thought permissible.

It was not thus in ancient times. The earlier men were hunting men, and to hunt a neighboring tribe, kill the males, loot the village and possess the females, was the most profitable, as well as the most exciting, way of living. Thus were the more martial tribes selected, and in chiefs and people a pure pugnacity and love of glory came to mingle with the more fundamental appetite for plunder.

Modern war is so expensive that we feel trade to be a better avenue to plunder; but modern man inherits all the innate pugnacity and all the love of glory of his ancestors. Showing war's irrationality and horror is of no effect upon him. The horrors make the fascination. War is the *strong* life; it is life *in extremis*; war-taxes are the only ones men never hesitate to pay, as the budgets of all nations show us.

History is a bath of blood. The Iliad is one long recital of how Diomedes and Ajax, Sarpedon and Hector *killed.* No detail of the wounds they made is spared us, and the Greek mind fed upon the story. Greek history is a panorama of jingoism and imperialism—war for war's sake, all the citizens being warriors. It is horrible reading, because of the irrationality of it all—save for the purpose of making "history"—and the history is that of the utter ruin of a civilization in intellectual respects perhaps the highest the earth has ever seen.

Those wars were purely piratical. Pride, gold, women, slaves, excitement, were their only motives. In the Peloponnesian war for example, the Athenians ask the inhabitants of Melos (the island where the "Venus of Milo" was found), hitherto neutral, to own their lordship. The envoys meet, and hold a debate which Thucydides gives in full, and which, for sweet reasonableness of form, would have satisfied Matthew Arnold. "The powerful exact what they can," said the Athenians, "and the weak grant what they must." When the Meleans say that sooner than be slaves they will appeal to the gods, the Athenians reply: "Of the gods we believe and of men we know that, by a law of their nature, wherever they can rule they will. This law was not made by us, and we are not the first to have acted upon it; we did but inherit it, and we know that you and all mankind, if you were as strong as we are, would do as we do. So much for the gods; we have told you why we expect to stand as high in their good opinion as you." Well, the Meleans still refused, and their town was taken. "The Athenians," Thucydides quietly says, "thereupon put to death all who were of military age and made slaves of the women and children. They then colonized the island, sending thither five hundred settlers of their own."

Alexander's career was piracy pure and simple, nothing but an orgy of power and plunder, made romantic by the character of the hero. There was no rational principle in it, and the moment he died his generals and governors attacked one another. The cruelty of those times is incredible. When Rome finally conquered Greece, Paulus Aemilius was told by the Roman Senate to reward his soldiers for their toil by "giving" them the old kingdom of Epirus. They sacked seventy cities and carried off a hundred and fifty thousand inhabitants as slaves. How many they killed I know not; but in Etolia they killed all the senators, five hundred and fifty in number. Brutus was "the noblest Roman of them all," but to reanimate his soldiers on the eve of Philippi he similarly promises to give them the cities of Sparta and Thessalonica to ravage, if they win the fight.

Such was the gory nurse that trained societies to cohesiveness. We inherit the warlike type; and for most of the capacities of heroism that the human race is full of we have to thank this cruel history.

Dead men tell no tales, and if there were any tribes of other type than this they have left no survivors. Our ancestors have bred pugnacity into our bone and marrow, and thousands of years of peace won't breed it out of us. The popular imagination fairly fattens on the thought of wars. Let public opinion once reach a certain fighting pitch, and no ruler can withstand it. In the Boer war both governments began with bluff but couldn't stay there, the military tension was too much for them. In 1898 our people had read the word "war" in letters three inches high for three months in every newspaper. The pliant politician McKinley was swept away by their eagerness, and our squalid war with Spain became a necessity.

At the present day [1910], civilized opinion is a curious mental mixture. The military instincts and ideals are as strong as ever, but are confronted by reflective criticisms which sorely curb their ancient freedom. Innumerable writers are showing up the bestial side of military service. Pure loot and mastery seem no longer morally avowable motives, and pretexts must be found for attributing them solely to the enemy. England and we, our army and navy authorities repeat without ceasing, arm solely for "peace," Germany and Japan it is who are bent on loot and glory. "Peace" in military mouths today is a synonym for "war expected." The word has become a pure provocative, and no government wishing peace sincerely should allow it ever to be printed in a newspaper. Every up-to-date dictionary should say that "peace" and "war" mean the same thing, now *in posse* [potentially], now *in actu* [actually]. It may even reasonably be said that the intensely sharp competitive *preparation* for war by the nations *is the real war*, permanent, unceasing; and that the battles are only a sort of public verification of the mastery gained during the "peace"-interval.

It is plain that on this subject civilized man has developed a sort of double personality. If we take European nations, no legitimate interest of any one of them would seem to justify the tremendous destructions which a war to compass it would necessarily entail. It would seem as though common sense and reason ought to find a way to reach agreement in every conflict of honest interests. I myself think it our bounden duty to believe in such international rationality as possible. But, as things stand, I see how desperately hard it is to bring the peace-party and the war-party together, and I believe that the difficulty is due to certain deficiencies in the program of pacificism which set the militarist imagination strongly, and to a certain extent justifiably, against it. In the whole discussion both sides are on imaginative and sentimental ground. It is but one utopia against another, and everything one says must be abstract and hypothetical. Subject to this criticism and caution, I will try to characterize in abstract strokes the opposite imaginative forces, and point out what to my own very fallible mind seems the best utopian hypothesis, the most promising line of conciliation.

In my remarks, pacificist though I am, I will refuse to speak of the bestial side of the war-*régime* (already done justice to by many writers) and consider only the higher aspects of militaristic sentiment. Patriotism no one thinks discreditable; nor does any one deny that war is the romance of history. But inordinate ambitions are the soul of every patriotism, and the possibility of violent death the soul of all romance. The militarily patriotic and romantic-minded everywhere, and especially the professional military class, refuse to admit for a moment that war may be a transitory phenomenon in social evolution. The notion of a sheep's paradise like that revolts, they say, our higher imagination. Where then would be the steeps of life? If war had ever stopped, we should have to re-invent it, on this view, to redeem life from flat degeneration.

Reflective apologists for war at the present day all take it religiously. It is a sort of sacrament. Its profits are to the vanquished as well as to the victor; and quite apart from any question of profit, it is an absolute good, we are told, for it is human nature at its highest dynamic. Its "horrors" are a cheap price to pay for rescue from the only alternative supposed, of a world of clerks and teachers, of co-education and zo-ophily [opposition to vivesection], of "consumer's leagues" and "associated charities," of industrialism unlimited, and femininism unabashed. No scorn, no hardness, no valor any more! Fie upon such a cattleyard of a planet!

So far as the central essence of this feeling goes, no healthy minded person, it seems to me, can help to some degree partaking of it. Militarism is the great preserver of our ideals of hardihood, and human life with no use for hardihood would be contemptible. Without risks or prizes for the darer, history would be insipid indeed; and there is a type of military character which every one feels that the race should never cease to breed, for every one is sensitive to its superiority. The duty is incumbent on mankind, of keeping military characters in stock—of keeping them, if not for use, then as ends in themselves and as pure pieces of perfection—so that [Theodore] Roosevelt's weaklings and mollycoddles may not end by making everything else disappear from the face of nature.

This natural sort of feeling forms, I think, the innermost soul of army-writings. Without any exception known to me, militarist authors take a highly mystical view of their subject, and regard war as a biological or sociological necessity, uncontrolled by ordinary psychological checks and motives. When the time of development is ripe the war must come, reason or no reason, for the justifications pleaded are invariably fictitious. War is, in short, a permanent human *obligation*. General Homer Lea, in his recent book *The Valor of Ignorance*, plants himself squarely on this ground. Readiness for war is for him the essence of nationality, and ability in it the supreme measure of the health of nations.

Nations, General Lea says, are never stationary—they must necessarily expand or shrink, according to their vitality or decrepitude. Japan now is culminating; and by the fatal law in question it is impossible that her statesmen should not long since have entered, with extraordinary foresight, upon a vast policy of conquest—the game in which the first moves were her wars with China and Russia and her treaty with England, and of which the final objective is the capture of the Philippines, the Hawaiian Islands, Alaska, and the whole of our Coast west of the Sierra Passes. [Just six years before James wrote, Japan attacked and defeated Russian forces at Port Arthur, Mukden and Tsushima.] This will give Japan what her ineluctable vocation as a state absolute forces her to claim, the possession of the entire Pacific Ocean; and to oppose these deep designs we Americans have, according to our author, nothing but our conceit, our ignorance, our commercialism, our corruption, and our feminism. General Lea makes a minute technical comparison of the military strength which we at present could oppose to the strength of Japan, and concludes that the islands, Alaska, Oregon, and Southern California would fall almost without resistance, that San Francisco must surrender in a fortnight to a Japanese investment [seige], that in three or four months the war would be over, and our republic, unable to regain what it had heedlessly neglected to protect sufficiently, would then "disintegrate," until perhaps some Caesar should arise to weld us again into a nation.

A dismal forecast indeed! Yet not unplausible, if the mentality of Japan's statesmen be of the Caesarian type of which history shows so many examples, and which is all that General Lea seems able to imagine. But there is no reason to think that women can no longer be the mothers of Napoleonic or Alexandrian characters; and if these come in Japan and find their opportunity, just such surprises as *The Valor of Ignorance* paints may lurk in ambush for us. [This was written 31 years before Pearl Harbor.] Ignorant as we still are of the innermost recesses of Japanese mentality, we may be foolhardy to disregard such possibilities.

Other militarists are more complex and more moral in their considerations. The *Philosophie des Krieges* [*The Philosophy of War*], by S. R. Steinmetz is a good example. War, according to this author, is an ordeal instituted by God, who weighs the nations in its balance. It is the essential form of the State, and the only function in which peoples can employ all their powers at once and convergently. No victory is possible save as the resultant of a totality of virtues, no defeat for which some vice or weakness is not responsible. Fidelity, cohesiveness, tenacity, heroism, conscience, education, inventiveness, economy, wealth, physical health and vigor—there isn't a moral or intellectual point of superiority that doesn't tell, when God holds his assizes and hurls the peoples upon one another.

The virtues that prevail, it must be noted, are virtues anyhow, superiorities that count in peaceful as well as in military competition; but the strain on them, being infinitely intenser in the latter case, makes war infinitely more searching as a trial. No ordeal is comparable to its winnowings. Its dread hammer is the welder of men into cohesive states, and nowhere but in such states can human nature adequately develop its capacity. The only alternative is "degeneration."

Dr. Steinmetz is a conscientious thinker, and his book, short as it is, takes much into account. Its upshot can, it seems to me, be summed up in Simon Patten's word, that mankind was nursed in pain and fear, and that the transition to a "pleasure-economy" may be fatal to a being wielding no powers of defence against its disintegrative influences. If we speak of the *fear of emancipation from the fear-régime*, we put the whole situation into a single phrase; fear regarding ourselves now taking the place of the ancient fear of the enemy.

Turn the fear over as I will in my mind, it all seems to lead back to two unwillingnesses of the imagination, one aesthetic, and the other moral; unwillingness, first to envisage a future in which army-life, with its many elements of charm, shall be forever impossible, and in which the destinies of peoples shall nevermore be decided quickly, thrillingly, and tragically, by force, but only gradually and insipidly by "evolution"; and, secondly, unwillingness to see the supreme theatre of human strenuousness closed, and the splendid military aptitudes of men doomed to keep always in a state of latency and never show themselves in action. These insistent unwillingnesses, no less than other aesthetic and ethical insistencies, have, it seems to me, to be listened to and respected. One cannot meet them effectively by mere counter-insistency on war's expensiveness and horror. The horror makes the thrill; and when the question is of getting the extremest and supremest out of human nature, talk of expense sounds ignominious. The weakness of so much merely negative criticism is evident—pacificism makes no converts from the military party. The military party denies neither the bestiality nor the horror, nor the expense; it only says that these things tell but half the story. It only says that war is *worth* them; that, taking human nature as a whole, its wars are its best protection against its weaker and more cowardly self, and that mankind cannot *afford* to adopt a peace-economy.

Pacificists ought to enter more deeply into the aesthetical and ethical point of view of their opponents. Do that first in any controversy, says J. J. Chapman, *then move the point*, and your opponent will follow. So long as anti-militarists propose no substitute for war's disciplinary function, no *moral equivalent* of war, analogous, as one might say, to the mechanical equivalent of heat, so long they fail to realize the full inwardness of the situation. And as a rule they do fail. The duties, penalties, and sanctions pictured in the utopias they paint are all too weak and tame to touch the military-minded. Tolstoï's pacificism is the only exception to this rule, for it is profoundly pessimistic as regards all this world's values, and makes the fear of the Lord furnish the moral spur provided elsewhere by the fear of the enemy. But our socialistic peace-advocates all believe absolutely in this world's values; and instead of the fear of the Lord and the fear of the enemy, the only fear they reckon with is the fear of poverty if one be lazy. This weakness pervades all the socialistic literature with which I am acquainted. Even in Lowes Dickinson's exquisite dialogue, high wages and short hours are the only forces invoked for overcoming man's distaste for repulsive kinds of labor. Meanwhile men at large still live as they always have lived, under a pain-and-fear economy—for those of us who live in an ease-economy are but an island in the stormy ocean—and the whole atmosphere of present-day utopian literature tastes mawkish and dishwatery to people who still keep a sense for life's more bitter flavors. It suggests, in truth, ubiquitous inferiority.

Inferiority is always with us, and merciless scorn of it is the keynote of the military temper. "Dogs, would you live forever?" shouted Frederick the Great. "Yes," say our utopians, "let us live forever, and raise our level gradually." The best thing about our "inferiors" today is that they are as tough as nails, and physically and morally almost as insensitive. Utopianism would see them soft and squeamish, while militarism would keep their callousness, but transfigure it into a meritorious characteristic, needed by "the service," and redeemed by that from the suspicion of inferiority. All the qualities of a man acquire dignity when he knows that the service of the collectivity that owns him needs them. If proud of the collectivity, his own pride rises in proportion. No collectivity is like an army for nourishing such pride; but it has to be confessed that the only sentiment which the image of pacific cosmopolitan industrialism is capable of arousing in countless worthy breasts is shame at the idea of belonging to *such* a collectivity. It is obvious that the United States of America as they exist today impress a mind like General Lea's as so much human blubber. Where is the sharpness and precipitousness, the contempt for life, whether one's own, or another's? Where is the savage "yes" and "no," the unconditional duty? Where is the conscription? Where is the blood-tax? Where is anything that one feels honored by belonging to?

Having said thus much in preparation, I will now confess my own utopia. I devoutly believe in the reign of peace and in the gradual advent of some sort of a socialistic equilibrium. The fatalistic view of the war-function is to me nonsense, for I know that war-making is due to definite motives and subject to prudential checks and reasonable criticisms, just like any other form of enterprise. And when whole nations are the armies, and the science of destruction vies in intellectual refinement with the sciences of production, I see that war becomes absurd and impossible from its own monstrosity. Extravagant ambitions will have to be replaced by reasonable claims, and nations must make common cause against them. I see no reason why all this should not apply to yellow as well as to white countries, and I look forward to a future when acts of war shall be formally outlawed as between civilized peoples.

All these beliefs of mine put me squarely into the antimilitarist party. But I do not believe that peace either ought to be or will be permanent on this globe, unless the states pacifically organized preserve some of the old elements of army-discipline. A permanently successful peace-economy cannot be a simple pleasure-economy. In the more or less socialistic future towards which mankind seems drifting we must still subject ourselves collectively to those severities which answer to our real position upon this only partly hospitable globe. We must make new energies and hardihoods continue the manliness to which the military mind so faithfully clings. Martial virtues must be the enduring cement; intrepidity, contempt of softness, surrender of private interest, obedience to command, must still remain the rock upon which states are built—unless, indeed, we wish for dangerous reactions against commonwealths fit only for contempt, and liable to invite attack whenever a center of crystallization for military-minded enterprise gets formed anywhere in their neighborhood.

The war-party is assuredly right in affirming and reaffirming that the martial virtues, although originally gained by the race through war, are absolute and permanent human goods. Patriotic pride and ambition in their military form are, after all, only specifications of a more general competitive passion. They are its first form, but that is no reason for supposing them to be its last form. Men now are proud of belonging to a conquering nation, and without a murmur they lay down their persons and their wealth, if by so doing they may fend off subjection. But who can be sure that *other aspects of one's country* may not, with time and education and suggestion enough, come to be regarded with similarly effective feelings of pride and shame? Why should men not some day feel that it is worth a blood-tax to belong to a collectivity superior in *any* ideal respect? Why should they not blush with indignant shame if the community that owns them is vile in any way whatsoever? Individuals, daily more numerous, now feel this civic passion. It is only a question of blowing on the spark till the whole population gets incandescent, and on the ruins of the old morals of military honor, a stable system of morals of civic honor builds itself up. What the whole community comes to believe in grasps the individual as in a vise. The war-function has grasped us so far; but constructive interests may some day seem no less imperative, and impose on the individual a hardly lighter burden.

Let me illustrate my idea more concretely. There is nothing to make one indignant in the mere fact that life is hard, that men should toil and suffer pain. The planetary conditions once for all are such, and we can stand it. But that so many men, by mere accidents of birth and opportunity, should have a life of *nothing else* but toil and pain and hardness and inferiority imposed upon them, should have *no* vacation, while others natively no more deserving never get any taste of this campaigning life at all—*this* is capable of arousing indignation in reflective minds. It may end by seeming shameful to all of us that some of us have nothing but campaigning [struggling], and others nothing but unmanly ease. If now—and this is my idea—there were, instead of military conscription a conscription of the whole youthful population to form for a certain number of years a part of the army enlisted against *Nature*, the injustice would tend to be evened out, and numerous other goods to the commonwealth would follow. The military ideals of hardihood and discipline would be wrought into the growing fiber of the people; no one would remain blind as the luxurious classes now are blind, to man's relations to the globe he lives on, and to the permanently sour and hard foundations of his higher life. To coal and iron mines, to freight trains, to fishing fleets in December, to dish-washing, clothes-washing, and window-washing, to road-building and tunnel-making, to foundries and stoke-holes, and to the frames of skyscrapers, would our gilded youths be drafted off, according to their choice, to get the childishness knocked out of them and to come back into society with healthier sympathies and soberer ideas. They would have paid their blood-tax, done their own part in the immemorial human warfare against nature; they would tread the earth more proudly, the women would value them more highly, they would be better fathers and teachers of the following generation.

Such a conscription, with the state of public opinion that would have required it, and the many moral fruits it would bear, would preserve in the midst of a pacific civilization the manly virtues which the military party is so afraid of seeing disappear in peace. We should get toughness without callousness, authority with as little criminal cruelty as possible, and painful work done cheerily because the duty is temporary, and threatens not, as now, to degrade the whole remainder of one's life. I spoke of the "moral equivalent" of war. So far, war has been the only force that can discipline a whole community, and until an equivalent discipline is organized, I believe that war must have its way. But I have no serious doubt that the ordinary prides and shames of social man, once developed to a certain intensity, are capable of organizing such a moral equivalent as I have sketched, or some other just as effective for preserving manliness of type. It is but a question of time, of skillful propagandism, and of opinion-making men seizing historic opportunities.

The martial type of character can be bred without war. Strenuous honor and disinterestedness abound elsewhere. Priests and medical men are in a fashion educated to it, and we should all feel some degree of it imperative if we were conscious of our work as an obligatory service to the state. We should be *owned*, as soldiers are by the army, and our pride would rise accordingly. We could be poor, then, without humiliation, as army officers now are. The only thing needed henceforward is to inflame the civic temper as past history has inflamed the military temper. H. G. Wells, as usual, sees the center of the situation. "In many ways," he says [in *First and Last Things*, 1908], "military organization is the most peaceful of activities. When the contemporary man steps from the street of clamorous insincere advertisement, push, adulteration, underselling and intermittent employment into the barrack-yard, he steps on to a higher social plane, into an atmosphere of service and coöperation and of infinitely more honorable emulations. Here at least men are not flung out of employment to degenerate because there is no immediate work for them to do. They are fed and drilled and trained for better services. Here at least a man is supposed to win promotion by self-forgetfulness and not by self-seeking. And beside the feeble and irregular endowment of research by commercialism, its little short-sighted snatches at profit by innovation and scientific economy, see how remarkable is the steady and rapid development of method and appliances in naval and military affairs! Nothing is more striking than to compare the progress of civil conveniences which has been left almost entirely to the trader, to the progress in military apparatus during the last few decades. The house-appliances of today, for example, are little better than they were fifty years ago. A house of today is still almost as ill-ventilated, badly heated by wasteful fires, clumsily arranged and furnished as the house of 1858. Houses a couple of hundred years old are still satisfactory places of residence, so little have our standards risen. But the rifle or battleship of fifty years ago was beyond all comparison inferior to those we possess; in power, in speed, in convenience alike. No one has a use now for such superannuated things."

Wells adds that he thinks that the conceptions of order and discipline, the tradition of service and devotion, of physical fitness, unstinted exertion, and universal responsibility, which universal military duty is now teaching European nations, will remain a permanent acquisition, when the last ammunition has been used in the fireworks that celebrate the final peace. I believe as he does. It would be simply preposterous if the only force that could work ideals of honor and standards of efficiency into English or American natures should be the fear of being killed by the Germans or Japanese. Great indeed is Fear; but it is not, as our military enthusiasts believe and try to make us believe, the only stimulus known for awakening the higher ranges of men's spiritual energy. The amount of alteration in public opinion which my utopia postulates is vastly less than the difference between the mentality of those black warriors who pursued Stanley's party on the Congo with their cannibal war-cry of "Meat! Meat!" and that of the "general staff" of any civilized nation. History has seen the latter interval bridged over: the former one can be bridged over much more easily.

On November 2, 1960, at the climax of his Presidential campaign, Senator John F. Kennedy proposed the creation of a Peace Corps, an idea which, upon taking office, he promptly put into effect. Kennedy was concerned not only with helping underdeveloped nations learn the skills they needed to help themselves, but also with counteracting the effects of Soviet agents trained, in his words, "to disrupt free institutions in the uncommitted world." Thus, the motives for starting the Peace Corps combined an idealistic melody with the familiar bass notes of anti-communism.

In accord with the Cold War side of Kennedy's vision, also expressed in his baseless campaign charge of a "missile gap" and in his stirring inaugural promise to "pay any price and bear any burden," the candidate urged that a Peace Corps would bolster U.S. prestige by outclassing Soviet technicians in the villages and barrios of the Third World. Of course, the volunteers would directly help the people among whom they would work abroad, but above all they would "serve our country."

In return for working abroad in this way as "ambassadors of peace," young male volunteers would probably be offered exemption from the draft, a delicate political move even in those peaceful times before the Vietnam War. In his prepared text, Kennedy described Peace Corps service as "an alternative to peacetime selective service," whereas in delivering the speech he added the words, "or as a supplement."*

Widely popular, the Peace Corps was among the first major achievements of the Kennedy Administration. At its numerical peak, in 1966, the Corps had over 15,000 volunteers; and even now, under Reagan, it has over 5,000, with an emphasis upon technical skills in fields such as agronomy, health care, engineering, and forestry.

All the Skills Necessary

John F. Kennedy

WHERE ARE WE GOING TO OBTAIN THE TECHNICIANS needed to work with the peoples of underdeveloped lands outside the normal diplomatic channels—and by "technicians" I include engineers,

*** This composite text of Kennedy's proposal includes in boldface the passages of his prepared text that he actually delivered; in roman type, the passages that he omitted from the original text; and in italics, some passages that he added as he spoke.**

doctors, teachers, agricultural experts, specialists in public law, labor, taxation, civil service—all the skills necessary to establish a viable economy, a stable government, and a decent standard of living?

A news item in this week's paper reported that "a group of Russian geologists, electrical engineers, architects, and farming and fishing experts arrived in Ghana today to give technical advice." Another item described the potentiality for a Castro-type of Communist exploitation in northeast Brazil, where intolerable living standards have reduced thousands to a starvation diet, and in two villages prevented a single baby from living beyond the age of 12 months. And still another item described unrest in the Caribbean island of Haiti, where 90 per cent of the population has never been taught to read or write, and there is one doctor for every 10,000 people.

Think of the wonders skilled American personnel could work, building goodwill, building the peace. There is not enough money in all America to relieve the misery of the underdeveloped world in a giant and endless soup kitchen. But there is enough know-how and enough knowledgeable people to help these nations help themselves.

I therefore propose that our inadequate efforts in this area be supplemented by a Peace Corps of talented young men *and women*, **willing and able to serve their country in this fashion for three years as an alternative** *or as a supplement* **to peacetime selective service—well-qualified through rigorous standards; well-trained in the language, skills, and customs they will need to know; and directed and paid for by [U.S. government foreign aid] agencies. We cannot discontinue training our young men as soldiers of war—but we also want them as "ambassadors of peace."** *The combat soldiers, like General Gavin who jumped with his division in northern France, said that no young man today could serve his country with more distinction than in this struggle for peace around the world.*

This would be a volunteer corps, and volunteers would be sought among not only talented young men and women, but all Americans, of whatever age, who wished to serve the great Republic and serve the cause of freedom; men who have taught, or engineers or doctors or nurses who have reached the age of retirement, or who in the midst of their work wish to serve their country and freedom, should be given an opportunity and an agency in which their talents could serve our country around the globe.

This nation is full of young people eager to serve the cause of peace in the most useful way. I have met them on campaigns across the country. When I suggested at the University of Michigan, lately, that we needed young men and women willing to give up a few years to serve their country in this fashion, the students proposed a new organization to promote such an effort. Others have indicated a similar response, offering a tremendous pool of talent that could work modern miracles for peace in dozens of underdeveloped nations. I am convinced that our young men and women, dedicated to freedom, are fully capable of overcoming the efforts of Mr. Khrushchev's missionaries who are dedicated to undermining that freedom.

Robert Fuller is a former President of Oberlin College. As an educator, he understands the value of learning by participation in another culture, such as in programs of extensive citizen exchanges. Imagine how the tone of American and Soviet life might be altered if, throughout each society, there were people of all classes (or levels of "classlessness") who, when some piece of routine nonsense was spoken about the other country, could simply say, "That isn't what *I* saw there—where I was, it was like this." For more of Fuller, see his classic interview, "A Better Game Than War."* Meanwhile, join him in imagining how a World Peace Corps could help to prevent war and enlarge minds.

Proposal for a World Peace Corps

Robert Fuller

THE U.S. PEACE CORPS HAS FUNCTIONED PRIMARILY ALONG the North-South axis—from America, unilaterally, to the "developing countries"—and is directed toward issues of resource sufficiency. The World Peace Corps would function primarily along the East-West axis, and instead of being unidirectional would be reciprocal in its citizen exchange. Its purpose is to enhance mutual security by increasing the understanding of, and tolerance for, societal differences.

The World Peace Corps is a constructive, participatory activity for citizens who want to work towards the prevention of war; a unifying national ideal, like the U.S. Peace Corps or the moon landing; a way to place active peacemaking efforts on the national agenda; and an inspiring message which could generate an outpouring of positive response to a presidential candidate, as did the original Peace Corps in the presidential campaign of 1960.

The proposal for a World Peace Corps envisions the U.S. Gov-

*** Reprinted in *Citizen Summitry*, edited by Don Carlson and Craig Comstock (Tarcher/St. Martin's Press, 1986).**

ernment taking the initiative to arrange exchanges of thousands of American, European, Soviet, Chinese and other citizens. In the beginning, these would probably be a set of bilateral arrangements among the nations participating. Eventually, the exchanges would be multilateral and, ideally, the organization would become an international rather than an American one.

The project would create a constructive channel for the groundswell of intense citizen concern with the prevention of war. As well, it would have the long-range effect of creating a body of Americans and citizens of other countries who have a substantial knowledge of, and tolerance for, other cultures. This in turn would exert an influence in the countries involved away from unrealistically fearful or dangerously ignorant international behaviors.

As everyone concerned about nuclear war knows, no longer can a person or nation do anything to enhance its own security separate from stabilizing the whole system (community, planet) of which it is a part. The World Peace Corps would encourage that larger perspective.

Behind this proposal lies the observation that wars in general, and the present arms race in particular, are made possible by the support and participation of citizens who fear the difference of the "other" (the "enemy"), and are led to fear his intentions. This distrust operates consistently in the justification of military activity.

In the past, it was primarily the elites of nations who were able to establish contact with their counterparts from other countries. A World Peace Corps would extend this opportunity to citizens at large. Thus it provides a counterbalance to one of the conditions necessary for sustaining wars and the preparations for war: excessive fear born of ignorance.

Imagine two working class families exchanging jobs and houses for a year—an autoworker in Detroit and his counterpart in the Soviet Union, for example—with each family committed to participating in the construction of mutual understanding. Imagine this among agricultural and industrial workers, and the caring professions—nurses, aides, doctors.

Selection of Peace Corps members could be made by the receiving countries, to fulfill specific needs. Each country would have placement offices.

Volunteers might have to forfeit their right to seek political asylum in the host country, for the plan to be politically feasible.

As in the U.S. Peace Corps, participants would undergo substantial training dealing with the problems of cultural difference, language barriers, and so forth.

Wouldn't this task be extremely complex and politically troublesome? Yes! It would be a tremendous challenge, perhaps as complex as many that are confronted in contemporary military planning. This is part of the point: to divert some of the enormous resources, energy and intelligence of these defense activities into a way of making peace with "software" instead of unstably maintaining it with "hardware."

The myriad crises that would undoubtedly arise in the course of such citizen exchanges would similarly be part of the benefits: better a thousand of these crises than one great catastrophe. Also, the preparation for living in a different culture and the difficulty in actually doing so for long periods of time would provide to participants many of the same personal and developmental challenges that war and preparation for war have offered in the past.

The entire project could be funded with less than one percent of current national defense budgets. Private sector money already being used for thousands of different educational exchanges could be coordinated with the World Peace Corps.

Already there are many people and institutions involved in cultural exchanges, citizen exchanges, and "track two diplomacy." Already there is a body of experience and skill in the U.S. Peace Corps as well as entities like the Foreign Service, ready to be called upon in support of such an effort. Already there is worldwide concern about nuclear war. Already there is a vast amount of money being poured through national governments for the prevention of war and the protection of national sovereignty.

At this moment the American citizenry approaches disillusionment with its government, and is ripe for an inspiring, creative leadership and for a chance to participate itself in building peace. There remains to be added only the clear intention, stated in the right words, at the right time, by a political leader ready to follow up with sincere action, toward the creation and success of the World Peace Corps.

Selected to study how the Army could elicit more of the unused potential of its members, Lieutenant-Colonel Jim Channon (ret.) presented his ideas by creating an imaginary "First Earth Battalion" and initiating some of his colleagues into it. In the process, he presented "over a hundred new ways of knowing, doing, and being that were being explored on the West Coast"—an experience that reportedly led to the Army's new recruiting slogan, "Be all you can be." Although some may feel that Channon's explorations are irrelevant to the Army's basic job, others will find that his creativity in reviving the ideal of the "warrior monk" will suggest a hundred ways not to *reject* military values but to *convert* them to socially beneficial tasks.

Channon drafted a field manual for his imaginary battalion, which was soon known to an estimated sixty percent of the middle-level officers in the U.S. Army. However, when the Army offered to start an actual unit called the First Earth Battalion, its inventors declined on the ground that the unit would assist in the assimilation of new ideas by the Army more if it were "mythical." If you think everything in the Army operates on the slogan, "If it moves, salute it, and if it doesn't, paint it," please read on.

The First Earth Battalion

Jim Channon

THIS IS A TRUE STORY. IT IS ABOUT THE UNITED STATES Army and some of the people in it. It covers a period from 1978 to the present. You will find this information surprising if you are at all like the others who have heard this story. It is a strange and troubling fact that so few people in this country understand the nature of their Army, the people in it, and most of all its capacity to serve them in a new way.

In 1978 the Army was, as usual, studying its capacity to perform on the battlefield. One command of the Army, called TRADOC, was interested in the combat effectiveness of the American battalion compared with a Soviet battalion in the 1985 time-frame. The work done by TRADOC revealed that by 1985 American and Soviet forces would be equipped with weapons and vehicles that were forecasted to be roughly equal in capacity. General Donn Starry, the commander of TRADOC, reckoned that since the weapons and equipment for 1985 were already designed and in production, *people* were the only way to make a clear and decisive difference. General Starry nominated Colonel Mike Malone to head up a task force to discover how to empower people. This task force, headed by the System's Doctrine Office at TRADOC, began to look at all the ways that soldiers could become more effective by themselves and in teams.

Mike Malone was a legend in the United States Army. He functioned pretty much like the Army's own Mark Twain. He had a way of writing about the Army and what it did that allowed the Army to see itself more clearly. Malone would spend a week with a rifle squad inside an armored personnel carrier in the desert during his vacation—that was his style. On assignment from the Pentagon he wrote stories about the quality of Army life in a very downhome, personal, and concerned way.

During his time in the Army, Mike Malone had made it his business to know of others who were innovators and who cared primarily for the lifestyle of soldiers. For this mission he collected a small group of specialists around him from all over the Army. They were quite an eclectic group, to say the least. And in their initial meetings they agreed on several strategies that they would use to uncover the information on human dimensions desired by TRADOC.

During its early months, in 1978–79, the group scanned a wide variety of disciplines. They looked at the living-systems work being done by the University of Louisville. They looked at trainings of peak performance done by the major practitioners in the country. They looked at traditional thinking on how humans function and operate, as in the adaptive coping cycle. And they even had a reporter working the West Coast to come up with the latest from the esoteric and New Age world. During the early months, this group became quite a family. It seems they understood that quality performance was something they must emulate as well as try to collect information on. Their think tank was a model of adventure into the science of performance. Soon up to a hundred and thirty others had joined the team, mostly by attraction.

This group came to be known as Task Force Delta. Borrowed from science, the Delta stood for the difference between where soldiers were now and their ultimate potential. And so, the purpose of the task force became to decrease the difference between capacity and potential and, working through people, to improve the performance of teams.

During the third meeting, one of the original selectees in the group, Lieutenant-Colonel Jim Channon, came in from the West Coast with his most recent collection of human potential technology, puzzled over how he might present his group of rather diverse experiences and tools. While flying to the Task Force Delta conference held at Fort Knox in the spring of 1979, he decided he would create a mythological Army unit called "The First Earth Battalion," within which he could include all these many disparate ideas and experiences.

One evening, on the third day of this conference, Channon invited the 80-some-odd participants to the officers' club at Fort Knox to be "initiated" into the First Earth Battalion. The group assembled included Army officers from captain to general, civilian scientists and a group of friendly civilian consultants who were so interested in the problem-solving style of this creative group that they donated their services to the Army at no charge.

When the group arrived at the officers' club that night they were asked to remove their shoes, as they worked their way into one of the largest sitting rooms in the back of the club. There, Channon had assembled a circle of chairs which were sandwiched between plants that had been brought in from all around the club. The room was darkened except for a candle in the center placed next to a dollar bill.

The group had heard something of the style of this officer. They had expectations that they might be in for something "different" and their expectations were justified, for Channon opened the evening ceremony about this mythical battalion in the following way:

"Suspend your thinking for the next 45 minutes. Then after experiencing this sequence of events apply your rational mind to them and let us decide whether we have anything worthy of exploration or not."

The group giggled nervously and took their seats. Channon first asked them to stand together in a large circle and intone "e" as a mantra or ritual sound. "E," he explained, "stood for Earth, the life support system of every soldier," and through this harmonious intoning, he explained, we could in fact align our being for the remainder of the ceremony.

He said, "Swallow your saliva, take a slow easy breath, fill your stomach and then easily and naturally expel the sound 'e' with me together."

They all did that, of course, because they were soldiers, and they followed orders. Nevertheless, the first sounds that came from their mouths were awkward, squeaky, and eventually spilled out into nervous laughter.

He explained to them quickly and straightforwardly that what they had done was not an acceptable performance. They dutifully swallowed their saliva and began again. They were soldiers. This time, their nervousness relieved, the sound "eeeee" flowed forth beautifully and harmoniously. It set the stage for the following ritual which they all received comfortably and gracefully.

After several more experiences standing together, the group was enjoined to take their seats around the circle where Channon then explained the mystical nature of the dollar bill. "This piece of communication," he began, "is the most widely printed piece of information on the planet Earth. It says, 'In God we trust.' One of the major symbols is the unfinished pyramid completed by the all-seeing eye." He translated the Latin roughly as "This is the order of the ages in which God is well pleased." On the other side of the bill he noted that the very strong image of the American eagle had been turned toward the arm holding the olive branch symbolizing peace. He mentioned that John F. Kennedy had done that only recently. "E pluribus unum," he explained, could mean "Unity empowered by diversity." Another way of saying it was "To value the difference."

He followed this general line of thinking with a description of a day in the life of the soldier of the First Earth Battalion. In this way he was able to include such interesting features as breathing techniques with other things as mundane as natural toothpaste. It was in the melting pot of the entire day's experience of this hypothetical soldier from this mythical unit that Channon was able to include over a hundred new ways of knowing, doing, and being that were being explored on the West Coast.

At the end, Channon asked the group to apply their thinking now on how any of these ideas could be of any value to the Army in the future. Of course he expected some critical remarks about things that did not seem to have all that much direct application. To be sure there were those things. There were a great number of other things for which it was not possible to determine the benefit to the Army in that night's sitting. But to his surprise and delight, the primary response came from those who were interested in dealing with the human spirit. "Workspirit" they came to call it.

It seemed that during the modernization of the Army, a great number of very personal rituals and other forms of group camaraderie had taken a lower priority in the soldier's day. These officers were concerned with tending the soldier's human spirit during unrewarding and difficult times. They reckoned that the First Earth Battalion was a great experimental vehicle to try out and test new forms of bonding, new forms of mythology, and new ways to affect positively the quality of life and the human spirit of the American soldier. And so, that evening, at Fort Knox, Kentucky, the First Earth Battalion was born.

Some very interesting things occurred in the three years that followed the introduction of the First Earth Battalion to the Army. The Delta Force and the First Earth Battalion became compatible projects. The Delta Force, during conferences, would examine systematically all the scientific and more formal aspects of human performance. During the evenings, sometimes late into the night and sometimes at nearby retreats, the First Earth Battalion would experiment with new ways of being. The First Earth Battalion flavor which was introduced into this rather traditional think tank gave it a feeling of family. Men embraced each other as only men who are confident of their masculinity can embrace each other. Many women were invited and voluntarily joined the organization. Of about 300 members at the peak of Delta Force and First Earth Battalion, almost one-third were women. This balance of male creativity and female intuition seemed to serve the purposes of the organization well.

Developing Institutions That Favor Peace

During the following years, Task Force Delta studied other questions such as leadership, readiness, and the creation of the Army's new high-tech light division. The First Earth Battalion meanwhile pioneered in collecting new values, new ways to be, and human performance tools from the most esoteric as well as practical disciplines. Insignia and slogans were developed, and all the elements of a mythical culture were brought to bear to make this experiment truly informative and fun to be a part of. Channon compiled a field manual from this mythical and hypothetical information. Task Force Delta distributed this to all the members of the task force in 1981. Today by rough count nearly 60 percent of the middle-grade officers in the United States Army have read, know about, or have their own personal copy of the First Earth Battalion field manual called *Evolutionary Tactics*. In addition to values and peak performance technology, the manual went into other interesting areas such as ethical combat, conflict illumination and eco-force operations.

These were new ways of prosecuting conflict in the information age and converting the military into its larger role of service—that of protecting the natural resources on the planet. These became clear options in the minds of all those officers who read this material. In short, it is safe to say that your Army has considered some alternative ways to serve. They are excited about serving in these ways, and many young officers have actually requested duty in this mythical battalion.

The Secretary of the Army, Marsh, asked that the First Earth Battalion be considered as an experimental unit in the Army. After some thought it was decided that the battalion itself wouldn't assist in the assimilation of new ideas as fast if it were real as it would if it were organized as a mythological unit. It seems that the Earth Battalion manual served as a shopping cart of ideas for officers all over the Army who felt free to use these ideas one at a time as they applied in their own circumstances. In this way, each of them felt that they were a part of the First Earth Battalion. Whereas, if there had been a real First Earth Battalion somewhere else, then they collectively wouldn't have been able to belong. Further, it was decided that fair evaluation would probably be difficult for a concept that was so extraordinarily different from the ones currently designed for the modern battlefield.

Some more information about the Army is in order. Did you know that four of the Army's top-ranking generals now all got to their positions because they were innovators? One was an innovator in training technology and included teaching and training techniques that far surpass anything used throughout the rest of the world today. We are talking about training simulators that involved the reality of combat as supported by computers, television, and other electronic media. One of the other generals was an innovator in management techniques, another an innovator in short-cutting the production and development cycle for new equipment, and another an expert in systems. So, despite the fact that the army has its share of bureaucratic diseases it manages to reward innovation.

West Point was created during the time of Thomas Jefferson. It was designed to give this country its first civil engineers in order to develop the frontier. (The Corps of Engineers is still responsible for the nation's water system which delineates bioregions based on watershed principles.) The bulk of the officers who end up in top leadership positions in the Army still come from this institution of civil engineering. Understand that from the beginning the Army's basic mission was to develop the frontier, and that in the future this large force of idealistic and service-oriented people could be put back to the task of restoring the nation's life-lines and natural resources.

Another surprise for you might be that the American Army and Navy have routine meetings with their Soviet military counterparts. The Navy meets monthly with their counterparts to see that no miscues on the sea could create unnecessary conflict. The Army has met with its counterparts along the German border, at the General Staff College in Fort Leavenworth, Kansas, and at Fort Lewis, Washington, one of its regular division posts. These visits have been returned in kind by our officers to their Soviet counterparts. Did you know that the commander of allied forces in Europe meets with his Soviet counterpart four days every year in one or the other's home? It is general knowledge that the Soviet Army has participated in many civic action projects on behalf of their people and their land. What's important about all this information is to know that there is a clear potential for the two Armies to meet together, agree on how to disarm the nuclear spectre, and then participate together in restoring their respective parts of the planet to their potential.

Perhaps this seems to you too much to take in one breath. What's suggested here would not be easily completed. However, the capacity for this kind of idealistic service exists, and it exists not only in our Army but in other armies of the world. As a member of the citizenry of this country and this planet you've chosen to have those who serve you perform in other ways. The American Army is not the palace guard of the American government and it never has been. The American Army is an instrument of the people of the United States and it supports by oath the Constitution of the United States. There is no one way to support and defend the Constitution or the people of the United States. Nuclear defense was a choice, but there are other choices.

For the sake of our children, we all deserve to look at more creative ways for our large institutions to serve us. There is great hope and promise that a large organization like the Army which we already value can be shifted to other work that both decreases the chance of nuclear war and increases the resources all of us might share into the coming century. This is high mission and elegant strategy.

I envision an international ideal of service awakening in an emerging class of people who are best called *evolutionaries* and who have soldier spirit within them. I see them come together in the name of *people* and *planet* to create a new environment of support for the positive growth of humankind and the living earth mother. Their mission is to protect the possible and nurture the potential. They are the evolutionary guardians who focus their loving protection and affirm their allegiance to people and planet for their own good and for the good of those they serve. I call them *ev*olutionaries, not *re*volutionaries, for they are potentialists, not pragmatists.

As their contribution to a hopeful future, these warrior monks bring evolutionary tactics. They recognize that the world community of peoples demands hope from those who would operate as servants of the people. Services rendered by the warriors of the First Earth Battalion are specifically designed to generate workable solutions to defuse the nuclear time bomb, promote international relations, spread wise energy use, enforce the ecological balance, assist wise technological expansion, and above all, stress human development.

Armies are both the potential instruments of our destruction and the organized service that can drive humanity's potential development. They are the "turn key" organizations that could either shift the energy of our world into a positive synergistic convergence, or bring us to the brink of the void. We have no choice but to encourage world armies to accept and express the nobility they already strive to attain. I can see their action expanding to include evolutionary work like planting vast new forests, completing large canal projects, helping in the design and construction of new energy-solvent towns, helping to clean up the inner cities, and working with the troubled inner city youth in young commando groups, and working harmoniously with other nations to see that the plentiful resources of our mother earth are equally shared by all peoples.

This will not be the first time that warrior monks have been active. In Vedic traditions, the warrior monk was a philosopher and teacher, and therefore a powerful transformational player. In the Chinese culture, the warrior was both a healer and teacher of martial arts. History affirms our own belief that there is no contradiction in the warrior and the service oriented monk prototypes living a completely harmonious, blended and parallel path when the basic ethic and service is "loving protection" of evolution and humankind. There is no contradiction in having armies of the world experience the same ethic as they evolve in peaceful cooperation towards the greater good of all.

It is sometimes difficult to determine how we have set ourselves against each other as nations, and even the more frustrating when we realize that the people of these nations are not really very different inside, and in fact have the same desires for growth and environmental balance and for prosperity that we have. But this is reality. And soldiers who have grown up in an "arms race" world are obviously doing their job of protection when they come up with a new and more effective weapons package.

But it is time for another approach, to use all of the military power for another end. It is time to give as much reward for the evolutionary contribution made by a soldier or an army as we have given in the past for the destructive contributions made on behalf of national defense.

I know that this process will begin with the transformation of soldiers and evolutionaries everywhere on the face of our planet home. There are young men and women who already aspire to this level of service and who are ready to make a permanent commitment. They will begin to meet in small groups to provide a support system for the personal transformation of group members. And on a small scale, these groups will begin selected evolutionary programs in their units and their communities.

All national level armies will begin to cooperate on ventures that stabilize the nuclear balance of terror. Joint teams could then patrol space and counter local terrorist activities that threaten stability in any given areas. Evolved cooperation will stifle the arms race.

Cooperation between the Soviets and American military could insure that neither side "flies off the handle" in direct collision in some local arena of tension, which otherwise could precipitate both sides into a major nuclear war.

As Channon proposes that warriors become servants of the earth, so Donald Keys describes how the United Nations, that haven of International civil servants, can sponsor military peacekeeping. As head of Planetary Citizens, Keys fosters global consciousness. Writing here as historian, he recounts the constitutional initiatives of the Secretary-General in enlarging the role of U.N. peacekeeping forces, especially during the Arab-Israeli war of October 1973.

As a result of these initiatives, the U.N. forces were "interposed" between the warring parties, allowed to defend themselves, supervised by the U.N. Secretariat, and paid out of general funds. A major improvement over previous rules for peacekeeping, all of this was successfully improvised under the most difficult circumstances. However, the U.N. still had to "pick up" forces for this role, the way a casual rock band might be assembled. And these forces could be interposed only with the consent of the two sides. Keys proposes that, before the next crisis emerges, U.N. members allow that body to develop a peacekeeping capability worthy of the name.

Peace-Keeping Forces

Donald Keys

IT HAS BEEN 40 YEARS SINCE THE UNITED NATIONS WAS founded by war-weary nations seeking a peace that would endure. This anniversary is a fitting time to take stock of the successes and failures in those years, to think about where the U.N. is going, and about where we would like to see it go. When nations can agree to act together in their common interest, much good can be accomplished. The U.N. has performed well in humanitarian areas, most often through the network of some 40 large international and intergovernmental organizations which are known as "the U.N. system." These include the Food and Agriculture Organization, the World Food Council, and the United Nations Environment Program as well as the World Health Organization which recently succeeded in abolishing smallpox from the planet.

Yet the U.N.'s record is disappointing as regards the primary purpose for which it was created: preserving the peace and developing alternatives to military conflict. Structurally, the U.N. is a pre-atomic organization, its Charter having been designed before the explosion of the Hiroshima bomb. The processes for securing peace inscribed in the Charter are therefore more lax and hopeful than they otherwise might have been. The optimistic idea of a concert of powers, the winners of World War II acting together to enforce peace, broke down almost immediately, and nothing has yet been designed to take its place. Although the U.N. has developed some powerful and effective peacekeeping tools for containing existing conflicts, their use has often been undermined by ambivalent mandates and a lack of resources. As for the planned alternative security system, it has never emerged from the embryonic stage. In these areas, the U.N.'s mandate has never been fulfilled because its Member States have been unwilling to give it the capacity and the mechanisms to be effective. Thus what we have had is not collective world security, but a series of stopgap and half measures, and therefore a mixture of successes and serious mishaps.

In fact, the distance between the state of the world at the end of World War II and the vision of the U.N. founders was too great to be bridged in a single step. The founders gave little guidance as to what specific actions might be taken to create "a new world order" from existing circumstances. Parts of that order have emerged of necessity over the years—usually in response to crisis—but much of the task still remains. Living as we do under the threat of nuclear destruction, it is a task that we *must* find a way to complete.

The process of building a U.N. which can truly preserve international security will require two interrelated parts. First, it will require a profound change in the way nations view their place in the world, and a reassessment of what is needed to provide security for their populations in the nuclear age. Many of the problems of the U.N.'s first 40 years make sense if we view them as reflections of tensions that naturally arise during transitions, when old paradigms that have structured our world view become outmoded, but new ones have not yet formed.

Second, building a United Nations worthy of the name will require a series of concrete steps to strengthen the system and gradually transform it into a true alternative security system. Without the existence of such an alternative system, our efforts at arms control are doomed to failure. Only when nations see global institutions that work will they be prepared to entrust their security to the world community.

There is an ironic circularity to the U.N. dilemma: nations, not believing it can work effectively in their interest, fail to give the U.N. the support it would need to perform its tasks; then they cite its weakness as a reason for their lack of support. To break out of this cycle of impotence, we propose a strategy of "confidence building" steps in the areas of U.N. peacekeeping and peacemaking—incremental changes that will noticeably improve U.N. functioning and build international support for further development. Such steps will be essential if we are to secure the willingness of nations to participate fully in building an alternative system strong enough to carry us through the next 40 years and beyond. So far the U.N. has been a function of its Member States' policies, basically reflecting their nationalistic foreign policies. As with arms reductions and disarmament, therefore, reforming the security functions of the U.N. must go hand in hand with the development of a greater sense of international political community. In the next sections, we will hear a rarely told story—how that development has taken place in the area of peacekeeping and we will suggest likely next steps.

In general, the U.N.'s peacekeeping forces have been consistently *undermanned, undermandated, underarmed* and *underfinanced*. Considering these handicaps it's remarkable what an important role they have played. The successful pursuit of U.N.-organized peacekeeping and peacemaking depends not only upon effective mechanisms within the world organization, but also upon the willingness of states to utilize them in good faith. As Charter signatories, states are committed to the notion of developing a legal world order in which they are bound to refrain from the use of force, and to the settlement of their disputes by peaceful means. All too often, of course, they fail to fulfill these obligations. Existing U.N. machinery for peaceful settlement of disputes remains rudimentary in the extreme, reflecting the lack of interest by nations in developing and using the U.N. for this truly central purpose. Any concrete acknowledgement and acceptance of the necessity for such procedures, normal and accepted in every national society but as yet novel in international affairs, will mark an important step in the maturation process of the global community.

A fundamental turning point in U.N. peacekeeping occurred in 1973. What peacekeeping arrangements had the U.N. evolved up to that point? In 1964–65, the U.N. was rocked by the so-called "Article 19 Crisis." Article 19 of the world organization's Charter states that Members which fall more than two years behind in their assessed payments to the U.N. shall lose the right to vote in the General Assembly. The U.S.S.R. and France had been withholding payments for peacekeeping operations with which they did not agree—which they were able to do, since such operations are funded by special assessment rather than from general membership payments. When the two-year threshold was reached, the U.S.S.R. insisted that the sanction did not apply to peacekeeping operations, and threatened to withdraw from the world organization if the matter were pressed. In an effort to avert a severe schism, the U.N. conducted all its business by consensus that session; needless to say, not much business was conducted. Out of that near disaster came the U.N.'s "Committee of 33"—the Special Committee on Peacekeeping Operations, which has been meeting ever since.

In 1970 the Committee began to register some progress on drawing up guidelines for "Model I" peacekeeping operations (in which the U.N. would be involved as an observer) and "Model II" operations (which would include the use of armed troops). This progress was somewhat illusory, however, as the Committee had left until last the thorny problems of control and financing. These issues remained unresolved. The U.S.S.R. wanted to assure tight Security Council control over any peacekeeping venture; the U.S. wanted to see the U.N. Secretariat continue to be the locus for day-to-day operations. A compromise was sought in the establishment of a subsidiary body of the Security Council, yet the superpowers could not agree on the proposed nature and functions of such a body. The essential impasse between the U.S. and U.S.S.R., especially over the relative roles of the Security Council and the Secretary-General, remained unbroken for two years, and some observers commented that there could be no agreement except as part of more general negotiations on questions of détente, in which the peacekeeping question might be a minor piece on a larger chessboard.

This stagnant state of affairs was unexpectedly shaken up in the Fall of 1973 by a crisis of world-threatening proportions. The October Arab-Israeli War erupted, bringing with it the danger of a superpower confrontation; the spectre of nuclear war loomed larger than at any time since the Cuban Missile Crisis. It is worth noting that the initiative for the Second United Nations Emergency Force (UNEF II) came not from any of the big powers, or Permanent Members of the Security Council, but from the nonaligned, nonpermanent members, and specifically from Yugoslavia. It was Yugoslavia's then U.N. Ambassador, Lazar Mojsov, who asked the Secretary-General about the feasibility of a new U.N. peacekeeping operation.

The Secretary-General replied that, by borrowing from the U.N. Force in Cyprus (UNFICYP), an initial U.N. presence could be on the ceasefire lines in the Sinai and Egypt within 24 hours. This led quickly to Security Council Resolution 340, offered by Yugoslavia, Guinea, India, Indonesia, Kenya, Panama, Peru, and the Sudan. It was a resolution which, under the circumstances, the U.S. and the U.S.S.R. were happy to accept, although not without some debate. Initially, the U.S.S.R. proposed that the two superpowers police the ceasefire; the U.S. insisted that the new force must be drawn from states other than the Permanent Members of the Security Council. Finally the U.S.S.R. agreed to accept the traditional "rule" that U.N. peacekeeping should be done by other than the major powers involved. China was persuaded by Egypt not to veto the resolution, and (in an innovative diplomatic stroke that allowed them not to vote "no") was listed "not participating." Under the main provisions of Resolution 340, the Security Council decided to set up a U.N. Emergency Force immediately. It affirmed that UNEF II would function under Security Council authority, and asked the Secretary-General to do three things: report within 24 hours on measures taken to implement the new Force; report on a continuing basis on the implementation of the resolution; and exclude Permanent Members of the Security Council from UNEF II. The Secretary-General agreed and at the same time asked for advance agreement to transfer troops from Cyprus to Egypt. No objections were raised. The fact that such forces could be transferred on short notice without at that time jeopardizing the situation in Cyprus was fortunate in the extreme, for it was to be weeks before UNEF II's own troop levels would be up to strength.

With innovative skill and speed the Secretary-General produced overnight the required report that was to provide the mandate and procedures for the new force. It is worth noting that the Security Council did not challenge the role of the Secretary-General in preparing for UNEF II, nor did it set up a subsidiary body of its own for the purpose. When the chips were down and the time pressure was on, the Council gladly relied on the accumulated expertise and experience of the Secretariat.

In one fell swoop the Secretary-General's ingeniously drafted "Report" of October 27, 1973 changed the face of transnational peacekeeping operations for the foreseeable future. The Report, approved with very little modification by the Security Council, firmly reasserted the traditional role of the genuinely international Secretariat in preparing and supervising U.N. peacekeeping operations, subject to the overall approval of the Security Council. The Secretariat was placed in charge of day-to-day operations, overseeing the Field Commander, and was given responsibility for negotiating troop and logistical contributions as well as agreements with "host" countries. The question of a subsidiary body, between the Security Council and the Secretary-General, did not arise. In this respect, the Report came closer to the U.S. than to the Soviet position in the Special Committee. Yet the Report also responded to an earlier Soviet concern that equitable geographic composition be considered in selecting countries from which contingents would be drawn; this was regarded by the Soviet Union as a major advance for its views.

A key precedent was set by Security Council approval of the Secretary-General's recommendation that the costs of the force were to be funded as part of the U.N.'s general expenses rather than by special payment. Since, as we have seen, trouble had already arisen over Members' defaults on special peacekeeping assessments, an integrated funding mechanism was crucial to providing a stable financial base for peacekeeping operations. Two entirely new features of the mandate, of the greatest potential importance, were established in other sections of the Report. The Self-Defense section provided that "The Force will be provided with weapons of a defensive character only. It shall not use force except in self-defense. Self-defense would include resistance to attempts by forceful means to prevent it from discharging its duties under the mandate of the Security Council." Thus, for the first time, a U.N. force was given authority actively to defend the territory for which it was responsible—to defend it, for example, from incursion by patrols preparing for an invasion across the buffer zone. Such authority may seem axiomatic to the casual observer, but it has not been granted in the past. While no one expects that the "thin blue line" of UNEF troops could hold a major invasion force at bay, their engagement in action could be a trip wire which, under the terms of the Report, would bring the Security Council into play.

A second major new provision was contained in the section on Continued Functioning: "All matters which may affect the nature of the continued effective functioning of the Force will be referred to the Council for its decision." This stipulation has been interpreted to mean that the Force cannot be removed from its task—for instance, at the request of the "host" countries—without the explicit approval of the Security Council. Of all the provisions of the approved Report, this is without doubt the most important, especially as regards the Middle East. One only need recall that the original UNEF was ordered ousted by the Egyptians in 1967, and this action was the immediate prelude to war. The U.N. troops were there purely on a "consent" basis, and when asked to go, their positions having already been overrun, they had no recourse. Thus, the U.N. presence, which had maintained an uneasy peace for ten years, was removed. If made part of future operations, both of the above mentioned provisions—for self-defense, and for Security Council action before any change is made in the status of the force—could provide insurance against a repetition of that unhappy event.

The mandate developed for UNEF II essentially short-circuited the work in the Committee of 33 by establishing precedents in the contended areas and in many more. This fact was not overlooked by the Committee. In its report to the 28th General Assembly it stated: "The current U.N. peace-keeping operations in the Middle East . . . are providing practical examples and constitute an experience which may assist the Special Committee and its Working Group in making further progress." The discussion of peacekeeping during the Assembly was much more fruitful and imaginative than it had been for many years.

In the wake of UNEF II, Members tried to nail down as precedent those parts of the mandate which particularly appealed to them and to dismiss the rest. One country's precedent was another's poison. The U.S.S.R. celebrated the precedent of the inclusion of an Eastern European country and equitable geographic distribution; the U.S. said it was no precedent. Within the Committee, the U.S. and its allies favored generalized guidelines rather than detailed ones, while the Soviet Union continued to press for a system which would avoid "unnecessary and sometimes dangerous improvisations."

That the mandate of UNEF II did in fact set precedents became evident when, in the wake of the Syria-Israeli agreement on disengagement in the Golan Heights area, the U.N. was asked to set up a U.N. Disengagement Observer Force (UNDOF). The Secretary-General's announced intention to "set up the Force on the basis of the same general principles as those defined" in his Report on establishment of UNEF II went unchallenged in the Security Council's authorization of UNDOF. In fact, some wording from the earlier Report found its way into the Protocol to the Agreement between Syria and Israel.

Building Confidence in Peacekeeping

Despite the progress made through UNEF II, peacekeeping missions continue to be undermanned, undermandated, and underfinanced. Further, the contrast between the experience in Egypt and in Cyprus demonstrates that, in the near future at least, U.N. peacekeeping can be successful only when the superpowers agree, and when it proceeds with the support of the Security Council.

UNEF II brought prestige to the U.N.; the U.N. Force in Cypress was to be cited as an example of U.N. ineffectiveness. In fact, however, the "defeat" of UNFICYP did not reflect a failure of U.N. skill but a failure of commitment on the part of member states. In a world crisis, nations were willing to move together; in a more localized conflict, they were unwilling to budge from their own agendas. Unfortunately, examples such as UNFICYP are held up as excuses for lack of recourse to the U.N. It is necessary therefore to take such steps as we can to increase U.N. effectiveness so as to build confidence in its viability as an alternative security system.

Articles 41 through 43 of the Charter explicitly mandate active U.N. intervention in cases that threaten the peace. But these enforcement provisions are too extreme for most situations, implying a U.N. "war" against transgressors. In contrast, peacekeeping processes, as thus far improvised, have been the opposite—too lenient and passive, allowing parties to a dispute to decide whether or not they want to receive U.N. observers or forces for interposition, and denying peacekeeping units effective or even reasonable mandates. Present mandates often jeopardize both the lives and the mission of U.N. units. There does not exist at present an understood or codified "middle ground" between relatively passive interposition with the permission of the disputing states on the one hand, and enforcement action as provided by the Charter on the other.

This is not to say that U.N. improvisations have not been valuable. In fact, "peacekeeping by interposition"—the placing of impartial troops to freeze conflict between adversaries—is a major contribution to the repertoire of an alternative security system. Indeed, it has much in common with the constabulary or police role in national societies. The weakness in this approach lies in the fact that it has been made dependent upon the consent of the disputing parties. Thus when Anwar Sadat decided to attempt to get Israel out of the Sinai, he was able to dismiss the U.N. forces which had been keeping the peace in the region. Although the mandate for UNEF II broke new ground in this area, stating that the operation could only be terminated by Security Council action, this precedent has not been institutionalized. Clearly, as long as peacekeeping operations are made and unmade at the whim of the nations involved, allowing them to freely pursue their narrowly perceived self-interest, the U.N. will not be able to do its job. It is therefore important that additional options be developed which lie between the U.N.'s acting as "policeman by invitation" and declaring war on aggressors. If it is truly to "secure and maintain peace" in the world, the U.N. will also require the option to interpose itself between adversaries, without their prior concurrence, when it has determined that a threat exists.

So far the U.N. has developed "pick-up" peacekeeping forces in response to crisis situations, but they are lamentably weak and underfinanced, and thus inadequate to the tasks they are called upon to do. Over the years, techniques for peacekeeping interventions have been developed, and the mandate for such operations has been tentatively extended. Yet to date, no Member of the Security Council has put forward or endorsed a comprehensive set of proposals designed to systematize what is now an ad hoc international peacekeeping capacity. Meanwhile the Secretary-General has warned in his *Annual Reports* that unless the Member States could quickly reform conflict resolution mechanisms and establish a new security system, the effectiveness of the U.N. would further deteriorate. Why wait until the next crisis to develop an adequate peace-keeping capability?

3 Redefining the Role of Business

In some "anti-war" circles it's widely assumed that all generals are "militaristic" and that people in business care about nothing but lucrative contracts, preferably from the Pentagon on a cost-plus basis. Evidence in favor of these assumptions is not difficult to find. However, our most eloquent warning against the effects of a permanent military establishment was delivered by a former general, and one of the most effective current critiques of excess military spending is coming from business executives.

In the U.S. the military sector of the economy has effects far beyond its percentage of the gross national product; and within the discretionary part of the Federal budget, the share assigned to the military is even larger than its share of the GNP. A vast lobbying operation on behalf of military spending has become a permanent part of Washington, promoting weaponry and opposing any foreign policies that might diminish the apparent need for higher levels of military hardware.

Drawing upon the public treasury, the military-industrial establishment competes effectively for the best talent in the areas of science, engineering, and manufacturing. If the U.S. has any industrial policy, it is set (or at least ratified) by the military, as in the current drive to recruit computer science on behalf of "Star Wars." (Meanwhile, of course, the Japanese are working toward new consumer products.) In turn, the jobs generated by military grants and orders create a large and well-placed constituency in support of the necessary funding. Asked to define an unbeatable weapon, a Pentagon colonel is said to have replied, "one that would employ subcontractors in every Congressional district."

Almost everyone agrees that we need a well-supported military. Some also specify that it ought to be efficient, capable of winning, organized by rational principles, and held in reasonable balance with other elements of the society that it protects. It was Dwight Eisenhower who cautioned a quarter century ago that the "military-industrial complex" was in danger of unbalancing the U.S. society and economy. As George Washington had warned about entangling foreign alliances, so in 1961 Eisenhower drew attention to the domestic political alliance that had recently coalesced in the favor of high and sustained military spending.

In the 1960s, I became aware that my father, an engineer and corporate excecutive, was helping to design the electrical power supply for missile silos. When I asked at the family dinner table how he felt about this work, he replied that the technical challenges were absorbing, his colleagues on the job were a cut above those he ordinarily encountered, and the U.S. had no choice whether to build the system. It was necessary to avoid a "missile gap" such as the one that Jack Kennedy had charged Eisenhower with allowing.

I understood his point, but I also suddenly realized that our Westchester County affluence was now being subsidized, in part, by the Pentagon. It was no longer "the government" building missile bases; it was my father. The sumptuous dinner that we were enjoying was bought by his skill in delivering the electrical power that would, if necessary, help to launch a missile. Should that ever occur, its warhead would explode, with the heat of the sun's surface, over some bull's-eye in the Soviet Union. On the other side of the North Pole, Soviet engineers were duplicating my father's work, in order to protect their motherland against the "imperialists."

At the family table we had been taught over the years to think in terms of the public good. Building underground silos was necessary because, without them, the other side would have nuclear supremacy and might attack us by surprise. My father said that. So, I'm sure, did his Soviet counterpart. Let's call him Nikolai. "Daddy," his son asks him in 1963, "why do we need so many strategic rockets?" "Son," Nikolai replies, "it's so the imperialists will never again be able to humiliate us as they did in Cuba." And so it goes—each daddy having the best of reasons as, year after year, he helps to construct the machinery of death.

Everyone will have his or her own story, many more intense or recent than mine—about a brother who designs guidance systems for rockets, a girl friend who programs computers for a defense contractor, a son who helps to assemble a fighter-bomber. There's a rationale for each of these projects. They pay well. Funds were voted by our elected representatives. Many peace-loving people share in the work. The Soviets are producing similar weapons, maybe more of them: "What would you do—let them get ahead?" (Or, as President Reagan would insist, *stay* ahead?) And, to an engineer, the projects are technically interesting.

At the time of the dinner conversation, I did not criticize my father for working on retaliatory forces (or, as the government likes to say, "defenses"). Actually, I shared his view that, in the absence of an agreement, they were necessary. But I did begin wondering more urgently about how the U.S. and the Soviet Union could agree to apply their resources and brains to other ends.

Many people are asking this question, often in surprising places. There's a cliché that the "peace movement" consists of people who are marginal to the "real world"—the sort of people who, in a shipwreck, are sent to the lifeboats first. In addition to the "women and children" (or students), there are marginal men such as religious leaders, scientists, other intellectuals, and "do-gooders," all of whom are thought to dwell on another plane of reality anyway.

It's not so easy, however, to dismiss the questions of business executives and TV anchors. During the Vietnam war, President Johnson knew he'd lost his support when the mainline media (represented, for example, by Walter Cronkite) and the financial and business community began to question the war in public, or simply to display its full reality. When CBS News ran a series filming a "search and destroy" mission, the President is reported to have said, "I've lost Walter." Anti-war ads signed by bankers, stockbrokers, and insurance executives began to appear in newspapers.

Something similar is starting to happen with regard to President Reagan's military buildup. In 1988, Republicans will surely have to answer the charges that they did not get very much for the money they spent on arms, that our economy has been disrupted in the process, that some of what they ordered is unnecessarily and dangerously provocative, and that our military forces are improperly organized to meet our commitments. I feel safe in making this prediction because the charges are already being prepared.[1]

It's one thing when such charges come from anti-war activists or the occasional academic analyst. It's another when one or more of these accusations are made by high retired officers (such as the former director of Advanced Space Programs Development for the Air Force), Presidential contenders (such as Gary Hart), and leading members of the business community (as in Business Executives for National Security, or BENS for short).

The business community is thought to favor "defense" spending automatically. It's said that building weapons stimulates the economy and creates spin-offs in the civilian sector, that it offers high profits. And anyway, what can executives, entrepreneurs, and other people in business contribute in areas that are traditionally left to military officers and politicians?

A bipartisan trade association, BENS has found much to say, and it is being heard.[2] Any Senator is going to pay careful attention upon being visited by a delegation including leaders of finance and industry from both parties, conservatives as well as liberals. In order to attract such a varied constituency and to focus its influence, BENS deals with a far narrower agenda than *Securing Our Planet* does. In this particular section, however, all the chapters are related to BENS.

Four of the contributors are members of it. Thomas J. Watson, Jr., and Regis McKenna both addressed a bi-coastal dinner celebration held by BENS in January 1986 to commemorate Dwight Eisenhower's Farewell Address of 1961. BENS has adopted this Farewell Address as a sort of charter, regarding Ike as if he were an honorary founding father of the organization. Among other contributors to this section, Don Carlson built the Western Region of BENS, and Harold Willens, a member of BENS, led a separate campaign in California for the nuclear freeze. While none of these people speak *for* BENS, their personal views suggest the energy and range of concerns that give the organization much of its power. A final chapter, by Lucien Rhodes, reports *on* BENS.

What's signified by the emergence of organizations such as BENS, and by the widespread critique of "Star Wars" on the part of many scientists as well as some retired officers, is that cracks are appearing in the military-industrial-scientific complex. I like to think that my father, if he were still active, would be part of this process. In these cracks new thoughts are growing. Here is a sample.

—Craig Comstock

1. See, for example, Edward N. Luttwak, *The Pentagon and the Art of War: The Question of Military Reform* (Institute for Contemporary Studies/Simon and Schuster, 1985); Gary Hart (with William S. Lind), *America Can Win: The Case for Military Reform* (Adler and Adler, 1986); and Robert M. Bowman, *Star Wars: A Defense Expert's Case Against the Strategic Defense Initiative* (Tarcher, 1986).
2. Business Executives for National Security maintains an office at the Euram Building, 21 Dupont Circle, N.W., Suite 401, Washington, D.C. 20036, and welcomes inquiries.

Like the Kennedy text reprinted in Section 1, Eisenhower's Farewell Address stands among the great recent Presidential speeches. In addition to warning against excessive power in the hands of military officers and of industrial leaders who profit by supplying weapons, Eisenhower also cautioned about the dynamic of scientific discovery and technical development—in particular, about the fantasy of creating "just one more" superweapon that would leave the other side totally at our mercy or, like "Star Wars," would defend us from the other side as if by magic.

Relying upon secret data from U-2 reconnaissance planes, Eisenhower held the line against excess military spending that was urged on the basis of Khrushchev's rhetoric rather than of actual Soviet weaponry.* As a former general, he had the authority to restrain the scaremongers. Thus, the following address represents not only a sentiment he allowed himself as he left the Presidency, but a belief upon which he acted while in office.

Farewell Address

Dwight D. Eisenhower

A VITAL ELEMENT IN KEEPING THE PEACE IS OUR MILITARY establishment. Our arms must be mighty, ready for instant action, so that no potential aggressor may be tempted to risk his own destruction.

Our military organization today bears little relation to that known by any of my predecessors in peacetime, or indeed by the fighting men of World War II or Korea.

Until the latest of our world conflicts, the United States had no armaments industry. American makers of plowshares could, with time and as required, make swords as well. But now we can no longer risk emergency improvisation of national defense; we have been compelled to create a permanent armaments industry of vast proportions. Added to this, three and a half million men and women are directly engaged in the defense establishment. We annually spend on military security more than the net income of all United States corporations.

*** Michael R. Beschloss, *May-Day: Eisenhower, Khrushchev and the U-2 Affair* (Harper & Row, 1986).**

This conjunction of an immense military establishment and a large arms industry is new in the American experience. The total influence—economic, political, even spiritual—is felt in every city, every State house, every office of the Federal government. We recognize the imperative need for this development. Yet we must not fail to comprehend its grave implications. Our toil, resources and livelihood are all involved; so is the very structure of our society.

In the councils of government, we must must guard against the acquisition of unwarranted influence, whether sought or unsought, by the military-industrial complex. The potential for the disastrous rise of misplaced power exists and will persist.

We must never let the weight of this combination endanger our liberties or democratic processes. We should take nothing for granted. Only an alert and knowledgeable citizenry can compel the proper meshing of the huge industrial and military machinery of defense with our peaceful methods and goals, so that security and liberty may prosper together.

Akin to , and largely responsible for the sweeping changes in our industrial-military posture, has been the technological revolution during recent decades.

In this revolution, research has become central; it also becomes more formalized, complex, and costly. A steadily increasing share is conducted for, by, or at the direction of, the Federal government.

Today, the solitary inventor, tinkering in his shop, has been overshadowed by task forces of scientists in laboratories and testing fields. In the same fashion, the free university, historically the fountainhead of free ideas and scientific discovery, has experienced a revolution in the conduct of research. Partly because of the huge costs involved, a government contract becomes virtually a substitute for intellectual curiosity. For every old blackboard there are now hundreds of new electronic computers.

The prospect of domination of the nation's scholars by Federal employment, project allocations, and the power of money is ever present—and is gravely to be regarded.

Yet in holding scientific research and discovery in respect, as we should, we must also be alert to the equal and opposite danger that public policy could itself become the captive of a scientific-technological elite.

It is the task of statesmanship to mold, to balance, and to integrate these and other forces, new and old, within the principles of our democratic system—ever aiming toward the supreme goals of our free society.

Another factor in maintaining balance involves the element of time. As we peer into society's future, we—you and I, and our government—must avoid the impulse to live only for today, plundering, for our own ease and convenience, the precious resources of tomorrow. We cannot mortgage the material assets of our grandchildren without risking the loss also of their political and spiritual heritage. We want democracy to survive for all generations to come, not to become the insolvent phantom of tomorrow.

Down the long lane of the history yet to be written America knows that this world of ours, ever growing smaller, must avoid becoming a community of dreadful fear and hate, and be, instead, a proud confederation of mutual trust and respect.

Such a confederation must be one of equals. The weakest must come to the conference table with the same confidence as do we, protected as we are by our moral, economic, and military strength. that table, though scarred by many past frustrations, cannot be abandoned for the certain agony of the battlefield.

Disarmament, with mutual honor and confidence, is a continuing imperative. Together we must learn how to compose differences, not with arms, but with intellect and decent purpose. Because this need is so sharp and apparent I confess that I lay down my official responsibilities in this field with a definite sense of disappointment. As one who has witnessed the horror and the lingering sadness of war—as one who knows that another war could utterly destroy this civilization which has been so slowly and painfully built over thousands of years—I wish I could say tonight that a lasting peace is in sight.

Happily, I can say that war has been avoided. Steady progress toward our ultimate goal has been made. But so much remains to be done. . . .

Redefining the Role of Business

Like several other contributors to this book—Don Carlson of Consolidated Capital, Regis McKenna, William Norris of Control Data, and Harold Willens—Thomas J. Watson is a businessman. In his case, the business is IBM, of which he was chairman of the board. Before reading his contribution, you may want to review the main points of Eisenhower's Farewell Address. The lessons that Watson draws from the address are clear, well illustrated, and richly relevant.

Eisenhower's Legacy

Thomas J. Watson, Jr.

IN JUNE 1944, I WAS SITTING IN A TENT IN SALALAH IN Oman pausing overnight while flying a group from Washington to India. Suddenly a soldier burst in and said, "The Normandy invasion has started!" Huddled over a shortwave radio, we listened to the details of the landings far into the night.

During that night, each of us in his own private way thought of the cold, lonely decision made twenty-four hours earlier by General Eisenhower.

Remember the coalition of some half dozen allied nations which had to be wooed and won by this remarkable man. Remember the last meeting: the General pointing to the weatherman and then to six or eight others for opinions, and then his two words, "We go." Remember how united all the allies were under his command. Persuasiveness, wisdom, diplomacy, courage, he had them all.

It was my good fortune to meet this great man in 1947 when my father persuaded him to come to Columbia University as its President. From that year until the time of his death, I had the opportunity of seeing him frequently and occasionally having a talk with him. The more I came to know this former military man, who became a veritable twentieth-century Man for All Seasons, the more I admired him.

We have all heard excerpts from his farewell address made on January 17, 1961. I should like to comment on three legacies I believe

President Eisenhower has left us—legacies reflected not only in the speech but throughout his career and particularly during his presidency: legacies which are as fresh and relevant today as they were twenty-five years ago.

The first legacy is realism—the cold realism of a man who looks facts in the face and acts on them.

Ike was no ideologue. As you know, he was very fond of bridge. In his public life also, he played the cards he was dealt. He dealt with the facts as they were and not as he and the American people wished they were. And he displayed cold realism again and again on many tough problems.

Some of you will recall the bloody battle of Dien Bien Phu when the French were finally defeated and forced out of Indochina. You will remember the calls that went out from France to President Eisenhower imploring us to join them in military action in the jungles of Southeast Asia. And you will remember his realistic and courageous position that this we should not do at any cost.

You will remember the bloody and inexcusable 1956 invasion of Hungary. You will remember the calls for help, the urging that we react. You'll recall how Ike and his Cabinet played the cards they were dealt. He recognized that the Soviet Union's area of influence—in effect decided at the end of World War II—included Hungary; and that in 1956 the Soviets had deliverable thermonuclear weapons, as we did. The President, again the realist, knew that we were in a new ball game. Painfully, he decided against intervention.

In these crises, President Eisenhower looked at the facts, bit the bullet, and did what he thought was right regardless of personal political consequences.

Most important of all, he realistically concluded more than thirty years ago that nuclear weapons had made another major war unthinkable; that the result would be the total destruction of western civilization. I draw the conclusion that President Eisenhower believed these weapons to be useful for only two things: deterrence or suicide.

We have been slow to recognize this fact. Things are much more hazardous than in Eisenhower's day. Sadly a recent poll found that over fifty percent of young Americans believe that their lives will be foreclosed by a nuclear war.

These weapons are infinitely powerful. One percent of either arsenal delivered on the other country would wipe out ninety cities. The facts are here for all to see: there is no viable defense. The Soviets will never let "essential equivalence" of nuclear forces disappear; nor will we.

Trust, approval, and endorsement of Soviet practices are not factors in this debate. No treaty should be made without certain verification, but we must realize that a fifty-fifty agreement is what we can get and treaties more favorable to us than to the U.S.S.R. are impossible to negotiate.

A recent Yankelovich poll reports that nearly ninety percent of the American people believe, as Ike did, that nuclear war would totally destroy both the United States and the Soviet Union. Recently, sentiment has been rising throughout the country again to try the treaty route. It was successful with the Limited Test Ban and with SALT I, and it is the only way to preserve the western world in the future.

Much of the news coming from the Congress and the Administration is confusing, however. On the one hand we hear that we are making a maximum effort at treaty-making, and on the other we hear it is impossible to consider the recent Soviet proposal of a Comprehensive Test Ban even with a provision for on-site inspections. Apparently, it is also impossible to consider treaties against deployment of weapons in space; and impossible to do numerous other things, all of which suggest that some political figures in Washington still think that nuclear arms can indeed be made useful.

We need to pay more attention to the legacy of realism.

The second legacy is balance.

The Military-Industrial Complex Speech is not a speech condemning the military-industrial complex. Ike liked and respected businessmen; thoroughly valued their contributions to the country; and recognized the necessity—for the first time in our history—of a large and efficient national arms industry.

But I believe he was concerned that this very important segment of our society would gain a momentum of its own, if uncontrolled by the Congress, the Administration, and ultimately the citizenry.

So in his farewell address, Eisenhower urged balance. He realized that national security depended on a three-fold balance—a strong military establishment, a strong economy, and a spiritual or patriotic strength in the people of the country to pull together.

President Eisenhower really told us there is no way to have a free ride. When I began to research how his fiscal responsibility turned out, I noted that he produced three out of the four balanced budgets the United States has had in the past thirty-five years.

How can we talk of strong defense against the Soviet threat and tolerate an annual deficit of $200 billion? It's time to return to balanced thinking.

The third Eisenhower legacy is a sense of ordered priorities. He knew how to put first things first.

For example, he was a Republican, and the names of some people who served with him have remained on the political scene still serving the Republican Party. But his greatness transcended party.

President Eisenhower believed in all America in a quiet, dignified, and courageous way. He believed that he was president of all the people.

He believed that as a nation, the United States had to compose differences with its allies, sometimes even subordinating our will to theirs. We cannot do it alone in time of peace any more than we did in time of war. And he believed what is good for the entire world—the entire human family—is also good for the United States.

Putting first things first, putting national good ahead of partisan advantage, President Eisenhower operated his Administration for eight years in the tradition of Truman, Marshall, Vandenberg, and many others who believed that bipartisan cooperation and consensus, particularly on national security, was vital.

He never played politics with military issues. Even during the 1960 campaign when facing the missile gap charge by his political opposition, he adamantly refused to take to the stump and release the classified information which would have refuted it.

Today, we debate all of our international problems in the public press and on television. Various members of the Senate, House, and the Administration express themselves freely, sometimes recklessly.

And the great historical non-partisan foreign policy of this country is all but lost and forgotten.

Above all, Eisenhower did not play politics with science. The "missile gap" disappeared shortly after President Kennedy took office. However, since that time, politicians have repeatedly used the Soviet Union, thermonuclear weapons, controversial technological possibilities, and perceived imbalances between the two countries as a means of gaining political strength.

If there is a single reason why we find ourselves close to an all-out arms race with the Soviet Union at the moment, the reason is that many politicians still hope, as so many of us did twenty-five years ago, that somehow our nuclear monopoly, which lasted only three or four years, can be repossessed by technological innovation.

In the light of these events, one sentence leaps out at us from Eisenhower's farewell speech: his warning against seeking a cure-all through "a huge increase in newer elements of our defense" or "a dramatic expansion in basic and applied research."

I don't know how President Eisenhower would have decided on Star Wars, the MX, or other current technological issues.

I believe I do know, however, how he would *go about* deciding. I believe he would seek the best scientific advice he could get: counsel from both sides of the aisle: counsel from the scientific community, not only from the hawks like Edward Teller, whose position and experience are useful, but from more moderate scientists like Jerome Wiesner, James Killian, Lee Du Bridge, Murray Gell-Mann, and others.

I believe that's what he would do, because that is exactly what he did when he was in the White House.

I am sure President Eisenhower realized that nuclear weapons would never be completely eliminated. The dangers of a nation's secretly developing a few and dominating the world are just too great. I surmise he did believe that treatying to lower numbers of weapons, equal on both sides, would reduce the possibility of a war by accident and reduce international tensions.

So there are three legacies—realism, balance, and ordered priorities.

And in the speech they all come together at a climactic moment—

not the section on the military-industrial complex, but the section at the end toward which the entire speech builds: a plea for disarmament; for the replacement of a "community of dreadful fear and hate" by a "proud confederation of mutual trust and respect"; a confederation in which "the weakest must come to the conference table with the same confidence" as the mightiest of nations—a conference table which, "though scarred by many past frustrations, cannot be abandoned for the certain agony of the battlefield."

That was his distant goal. His short-term goal was verifiable treaties to reduce the nuclear threat. That is why Gerard Smith, who negotiated SALT I, called President Eisenhower the father of arms control.

Dwight Eisenhower believed—genuinely believed—that the nations of the world in the long term could "learn how to compose differences not with arms, but with intellect and decent purpose."

That hope, I believe, is his greatest legacy of all. And he pledged that as long as he lived, he would never stop working to make that hope a reality, and he never did.

Eisenhower was one of history's greatest realists. In closing I quote his words written in April of 1956: "When we get to the point, as one day we will, that both sides know that in any outbreak of hostilities, regardless of the element of surprise, destruction will be both reciprocal and complete, possibly we will have enough sense to meet at the conference table with the understanding that *the era of armaments has ended* and that the human race must conform its actions to this truth or die."

In recent years there's been a major debate about whether the U.S. should, like Japan, have an "industrial policy." Proponents say that, in an age of fast-moving international competition, we can't go on depending solely upon the market to allocate investments. Opponents say that, however imperfectly the market may function, the government would do an even worse job. "Would you want IBM or Apple run by the Post Office?" they ask.

What got lost in this debate, for a while, is that the U.S. already has a de facto industrial policy. It's run by the Pentagon. It directs a substantial share of high technology industries into making what it wants, supplying investment, long-term orders, and even foreign customers. So the question is not whether America ought to have a deliberate policy for industrial redevelopment; it's whether we want it directed largely by the Pentagon.

Through his association with the founders of Apple Computer, as their primary marketing and public relations consultant, Regis McKenna helped to create Silicon Valley in its current form. In commenting on Eisenhower's Farewell Address, he pays special attention to the effect of excessive military spending upon our economic health, especially in the area of high technology.

Technology and Freedom

Regis McKenna

THE COMMON DENOMINATOR IN DWIGHT D. EISENHOWER'S life, in both his military service and in public office, was a devotion to the ideals of democracy. He not only fought for freedom against a monstrous external threat, but gave us his wisdom to protect that same freedom against a potential internal threat. On the 25th anniversary of his farewell address, let us reflect on how the freedom we all cherish is under attack—not by external military forces but by an encroaching weakness in our ability to compete economically.

We are now part of a highly competitive, global economy. During the brief era which began with our victory in the Second World War and ended with our defeat in Indochina, the crisis of Watergate and the emergence of OPEC, we came to believe that America was innately superior to the rest of the world due to our demonstrated prowess in technology and arms. We felt that our ability to innovate and produce at a pace ahead of any potential competitors was assured, and we turned our attention as a society to an ideological and geopolitical struggle with the Soviet Union. We were distracted. We lost sight of what was necessary to maintain the basis of our strength as a nation. The basis of America's strength is its industrial infrastructure and the ability to innovate, produce and compete. Without continued innovation and effective production, we cannot hope to compete in the new global economy.

America is bankrupting its commercial industries, hence its ability to compete, by spending disproportionate amounts of capital and human resources on the development and production of military weapons systems. But more than that, we are creating an economic incentive system, a Pentagon industrial policy, that directs the energies of our nation to nonproductive ends. Military spending should be regarded as a sacrifice to defend ourselves, but instead it has become an economic development program. But the role of the military is to defend this country, not to plan the economy.

To better understand the effect of our Pentagon industrial policy, consider what I term the "innovation cycle"—basic research, applied research, product development and production.

Basic research is performed by our universities, with much of the necessary funding from tax dollars. Indeed, the government plays a very important role in fundamental research. Because the government spends the bulk of its research dollars on military systems, the entire direction of product development and innovation is skewed away from the commercial sector.

The second phase of the cycle of innovation is applied research—research that transforms pure science into technology. It is often difficult to tell the difference between basic research and applied research, but in general applied research is directed more toward a specific end or a specific product, while basic research is directed more toward scientific discovery.

The product development stage is the point at which the microprocessor, the personal computer, recombinant DNA insulin, and other products all become working realities. It is in this stage that we see the prolific and productive results of entrepreneurism and venture capital. This nation has the most effective product development process in the world.

The production phase of the innovation cycle is the payoff in terms of return on social and private investment. This is the phase where new industries flourish and older industries are sustained. This is the phase that should achieve the social and economic return on years of investment in research and development. And it is this phase in which Japan, for one, has us beat.

While we still need to invest in research and development, we need to increase emphasis on the production end of the cycle if we are to remain globally competitive.

Where will we get the necessary funds and manpower to increase our competitive edge in research and development, and to bring us up to world competitiveness in production? Research and development spending in this country peaked in the mid-1960s at about 2.9 percent of the gross national product. Today we spend about $100 billion annually on R&D. That's about 2.4 percent of GNP.

But these numbers can be misleading. Over the past few years, there has been a pronounced shift in our spending patterns towards military-oriented development, specifically weapons development. The end products of this sort of research are of little value to our national strength. Missile maintenance creates jobs but it produces no new wealth to cycle back into the innovative process. An MX missile does nothing to enhance our competitive commercial posture. When we subtract military R&D, we find that the U.S. actually spends only 1.7 percent of GNP on non-military R&D, while Japan spends 2.3 percent and Germany 2.5 percent.

Unfortunately, the problem is worse still. For much of the federal funding for non-military R&D goes into space and energy research. I wonder how much of this sort of R&D is really helpful to our efforts to turn new technologies into globally competitive products. In fact, a study by the Library of Congress indicates that only 0.3 percent of federal R&D spending is directed toward the needs of commercial industry. In contrast, the comparable figure in Japan and Germany is 12.5 percent.

Funding military systems too heavily takes away from our economy. Besides creating a massive federal deficit that maintains high interest rates and staggering trade deficits from the resulting overvaluation of the dollar, it decreases the available resources for creating new technologies and commercial products. In addition, those engineers and scientists who would otherwise enter the commercial sector are wooed into military work by the lure of esoteric technology and unlimited budgets.

A weapons system, once completed, sits in a parking lot waiting for a war. It does not improve our standard of living, improve productivity or create opportunities for growth. Spend an equivalent amount on technological innovation in the commercial sector and you receive goods that increase America's productivity, competitiveness and opportunities for jobs and wealth creation. The result is a stronger America. Eisenhower said it well: "The military sector is not productive in and of itself, and if it takes too much of our total wealth, our national security is weakened."

Today, 70 percent of all federally funded research flows through military channels, which is a third of all research and development carried out in the United States. This figure does not include the $76 billion that will be spent on developing Star Wars before the Administration makes a decision on whether or not to deploy that system. The Defense Department is now the main supporter of scientific research in the U.S.

What of the theory that investing money in basic weapons research has civilian applications further down the line? Teflon, for example, initially developed for space and defense purposes, is used time and again to prove that military government spending results in consumer products. However, there are three holes in this argument. The first is that while military spending has increased, our competitiveness in commercial and industrial markets has dramatically declined. This is evidenced by the declining number of patents filed by U.S. companies as compared to the growing number filed by Japanese companies and those of other industrialized nations. The second hole in the argument is that only a few percent of all defense research dollars reach the consumer market as spinoff products. The third is that funding for military basic research, rather than specific weapons systems, has dropped dramatically.

This year, when the Department of Defense will spend more than $32 billion on research, development, evaluation and testing, only $861 million, or about 2 percent of the total, is to be spent on basic research which might be expected to further commercial technologies. Almost no commercial applications result from the development of particular weapons systems. The technologies developed for use in military systems are often too costly or sophisticated for commercial application.

The trickle-down theory—"What's good for the Pentagon is good for the country"—is further disproved when you look at the economy in Silicon Valley in the past two years. While military spending was at record levels, the U.S. semiconductor industry was in the worst recession of its thirty-year history and other members of the high-tech community suffered huge sales and profit losses. The only companies doing well are those in the military "old boys" club—those doing business directly with the Pentagon on a "cost plus" basis with no incentive to innovate at a low cost.

Another way that America is hurting its commercial industries through military spending is the wooing away of our best minds. Because the technical problems and challenges presented by Star Wars and the Strategic Computing Initiative are so fascinating, I fear that the best scientists and engineers will be drawn to defense work. This will cause other industries to fall behind in the international race. American Electronics Association chairman and former Undersecretary of the Navy William Perry has said that "the lack of electronic and computer science engineers may be the single most important factor limiting the growth and continued vitality of the electronics industries." If Star Wars is implemented as planned, it could consume all the electrical engineers who would graduate from California universities over the next ten years.

We need to ask ourselves if we can be at the cutting edge in the commercial sector while expending so much of our nation's talent on weapon systems development. Since World War II began, the U.S. defense establishment has absorbed between a third and a half of the nation's scientific talent. And given the way the Pentagon does business, those scientists and engineers are not being taught to minimize the cost of production, an area where we need much improvement.

The money spent on military systems over the past 25 years has led me to ask whether a permanent war economy has taken root, diverting talent and capital and thus crippling America's ability to compete with other industrial powers throughout the world, particularly Japan.

Like the United States, Japan has a large government bureacracy that is determining the direction of research and development. In this country, that bureaucracy is the Department of Defense, so the primary direction is toward military applications. In Japan the powerful governmental body that orchestrates industrial policy and allocates research dollars is the Ministry of International Trade and Industry, or MITI, and the direction is toward commercial economic development.

MITI targets which Japanese industries should grow and which industries and markets should be penetrated. It adjusts depreciation rates for favored industries and coordinates research into specific technologies. MITI is one reason Japan is dominating the consumer electronics semiconductor memory, machine tool and robotics markets. Research and development done under its auspices is partially responsible for Japan's recent and rapid success in the semiconductor industry. Because MITI, not a Department of Defense, is responsible for such direction, Japanese competitiveness and exports are particularly strong in commercial technology-based goods. The United States now has a growing trade deficit in technology-based products.

The Japanese investment in basic research is a relatively new activity. Previously, they let us do all the research and they entered the innovation cycle at the production end. Between 1950 and 1978, Japan acquired 32,000 new technology agreements, mostly from the U.S., at a bargain basement price of $9 billion. We spent, as a nation, some $500 billion developing that technology. Instead of investing in R&D, Japan invested in new factories, robotics, automation and international marketing. Instead of spending on the production of weapon systems, Japan spent its resources developing production systems that have brought American industry to its knees.

Now they own major markets, as we once owned markets, through high quality manufacturing. We simply haven't kept pace. Great product development is no longer enough. To compete, we need the capital and human resources for research, development and production that are now being directed in disproportionate amounts to the defense sector.

We readily acknowledge the need for a strong defense to protect America from all adversaries. Eisenhower's warning about the dangers of the "military-industrial complex" was not a call for its dissolution. It was an eloquent appeal for a balanced approach to national security based on our military, economic and intellectual resources. It was an appeal of wisdom from one who understood and cherished freedom. And it was an appeal that we not mortgage our freedom to a military-industrial economy. The direction of military spending, both in amounts and emphasis, is detrimental to our country's ability to compete economically. It actually weakens our national security.

Even though we have spent billions of dollars on defense, we have not won a war since 1945. What made the difference in World War II was our strong industrial base that could be quickly converted into the production of military supplies. The stronger our industrial base, the more secure we are. A weak American economy poses a threat to the American people and the world. All the weapons in the world will not safeguard a weak industrial economy.

I believe the answer is not in letting the defensive muscle of our country atrophy. Instead, it is in heeding President Eisenhower's call for balance between the commercial and military sectors, thus ensuring that freedom has its roots in a healthy economy.

Because both security and liberty stem in large part from economic security and economic liberty, let's put much more of our resources and efforts into strengthening America's ability to compete in global markets.

As Eisenhower so aptly said it 25 years ago: "Only an alert and knowledgeable citizenry can compel the proper meshing of the huge industrial and military machinery of defense with our peaceful methods and goals, so that security and liberty may prosper together."

Already introduced in Section 1, Harold Willens has written a book called *The Trimtab Factor,* explaining "how business executives can help solve the nuclear weapons crisis."* On the endpapers there's a chart in which all the firepower of World War II is symbolized by a single dot. The dot is surrounded by about 120 squares. Each of the squares contains about 50 dots. It takes all of these dots, which totally speck the endpapers, to represent the world's present nuclear weaponry. (An organization called Beyond War uses a sonic version of this chart, in which their speakers noisily drop an equivalent number of BB pellets into a can.) In this brief excerpt from his book, Willens explains the significance of its title, which he borrowed from Buckminster Fuller.

The Trimtab Factor

Harold Willens

I BELIEVE THE BUSINESS COMMUNITY HAS A UNIQUE ROLE to play in bringing about the transformation in consciousness that must occur if we are to survive. Business is the most flexible and change-oriented segment of our society. Constant interaction with marketplace realities makes it necessary for executives, managers, and entrepreneurs to reassess policies at regular intervals, evaluating them in terms of the costs, benefits, and risks involved. Successful business leaders must be ready to abandon unworkable policies and adopt new strategies to meet new challenges. This realistic and pragmatic approach must now be applied to our nation's nuclear weapons policy. The capabilities and characteristics that make for success in the competitive market-place are precisely those that can break the momentum of the nuclear arms race.

Over nearly forty years, the steady acceleration of the Soviet-American nuclear weapons race has given it a force and momentum that seem impossible to stop or change. Given the ideological, techno-logical, and bureaucratic inertia involved in our government's nuclear

*** Harold Willens, *The Trimtab Factor* (William Morrow, 1984).**

weapons policy, many Americans have come to believe that while the arms race is unfortunate, it is a necessary fact of life. But the continual accumulation of devices of mass destruction is not inevitable. We can assure our national security without continually adding to our nuclear arsenal. We can slow, stop, and gradually reverse the nuclear arms race.

In accomplishing this enormous task, the leverage of the business community can be a critical factor. Now, more than at any time in our history, we need to hear the voice of business, a voice that can make itself heard and heeded. Business leadership can provide what renowned architect, inventor, and philosopher R. Buckminster Fuller called the "trimtab" factor.

On airplane wings, and on the keels of racing yachts, trimtabs are small adjustable flaps that assist in balancing and steadying the motion of the craft. The principle of the trimtab also applies to a ship's rudder. In explaining the trimtab factor, Buckminster Fuller used the image of a large oceangoing ship traveling at high speed through the water. The mass and momentum of such a vessel are enormous, and great force is required to turn its rudder and change the ship's direction. In the past, some large ships had, at the trailing edge of the main rudder, another tiny rudder—the trimtab. By exerting a small amount of pressure, one person could easily turn the trimtab. The trimtab then turned the rudder, and the rudder turned the ship. Thus, the trimtab factor demonstrates how the precise application of a small amount of leverage can produce a powerful effect.

Business is the trimtab of America, a sector of society that possesses inordinate power to influence the direction of our national enterprise. Our ship of state is taking us into very dangerous waters. Business leadership can be the trimtab that will change our direction before it is too late.

Like McKenna, Carlson is concerned about the effects of military spending on our economic vitality. Often it's argued, both by defense industries and by their associated unions, that military spending is good for the economy. "In reality," says Carlson, "defense spending is a common sacrifice made for the common defense," and it should be justified solely upon that ground. Carlson speaks not only as a business executive, but as a resident of California, which receives over a fifth of all U.S. military contracts and which, after the Soviet Union and the rest of the U.S., would be the third largest military power on earth were it a nation. While many observers have assumed that these contracts are beneficial, Carlson, like Eisenhower, calls for balanced growth.

Security and the Bottom Line

Don Carlson

WE HAVE A SERIOUS PROBLEM. FOR TOO LONG, WE HAVE written blank checks to the Pentagon in the name of patriotism. We have accepted without question the fallacy that military spending is intrinsically good for the economy. Now, the hidden costs of military spending are eroding America's economic health and social well-being.

The Pentagon's $100 billion-per-year "military investment policy" is generating huge federal deficits, crowding out private-sector development, creating an internal "brain drain" that is hampering commercial innovation, undermining our competitive ability in international trade, and turning this country into a debtor nation. In the process, the "cost-plus" formula of Pentagon contracting has generated a vast military bureaucracy—the second largest planned economy outside of the Soviet Union—that is fraught with inefficiency, waste, and even fraud.

It is time to take an objective look at the real costs of military spending—from a bottom-line business perspective.

We often hear about the enormous share of the federal budget that goes for military expenses. If we hold taxes constant, we can say that this share competes with everything else that government could otherwise be doing to improve the national life; and if we allow for the possibility of tax reduction, we are reminded that the Pentagon competes directly with all of us as consumers and investors.

Many people complain about taxes, but despite the major role of the military in the federal budget, these complaints are nearly always deflected to other aspects of the budget. Very few public figures want to appear unwilling to give the military what it says it needs.

One exception was a man who could not be dismissed as antimilitary. Twenty-five years ago in his farewell address as President, Dwight Eisenhower warned that "this conjunction of an immense military establishment and a large arms industry is new in American experience." Alluding perhaps to the old Roman question about who will guard the guardians, Eisenhower declared that "in the councils of government we must guard against the acquisition of unwarranted influence, whether sought or unsought, by the military-industrial complex."

Unfortunately, not enough of us listened. Defense weaponry and military spending are not subjects that many of us have felt comfortable in studying or discussing in detail. We assumed that, like the economy, they are complicated subjects and better left to the experts. Unfortunately, many of those experts are in the peculiar position of being like the fox who is put in charge of the henyard. It's true they have experience but they are very prone to say "Why not?" to virtually everything in sight. Money is seldom of concern to anyone who is given a blank check.

Armament factories, research facilities, and military bases have spread across the United States. Similar proliferations have of course taken place in the Soviet Union, and on a smaller scale elsewhere. Hundreds of billions of dollars are involved. Tens of thousands of Americans, including a high percentage of our best scientists, are now employed in turning out an ever-increasing arsenal of nuclear weapons. (Over half of America's R&D brainpower is tied up on defense-related projects.)

Most of the other federal programs, including transportation, health, low income housing, national resources, and social services are being cut back while military expenditures have climbed. Well-meaning Congressional members have, for many years, sometimes found themselves in the unenviable position of not voting for or against a very dubious weapons system but, instead, for or against jobs in their congressional districts.

George Bernard Shaw once noted that "a government which robs Peter to pay Paul can always depend on the support of Paul." Paul symbolizes the military-industrial complex President Eisenhower warned us of and Peter is the vast majority of the rest of us, not to mention our children. This dole is being financed by deficit financing (government borrowing) at an unprecedented rate.

As we have come to learn, a military-industrial complex, whether here or in the Soviet Union, not only absorbs vast sums of money but also, through the dynamics of its own growth and of the technology it sponsors, has a major influence on foreign policy. Writing as a businessman, however, I want to address primarily the question of the economic effects of the military-industrial complex.

Although they are huge, the direct budget figures for military spending tell only part of the story. Whatever the federal government extracts through taxes and borrowing for use as "federal investments" is simply unavailable to private sector investors. The federal government's investments are also "de-coupled" from the marketplace. They don't promote the kind of business innovation and growth that is based on genuine market demand. The "growth" those federal investments produce is artificial growth, imposed from above; it's much less sustainable than market-based growth.

Federal investment can, of course, produce short-term bursts of economic activity, particularly in the defense industries, but those investments have two serious economic drawbacks: They produce far fewer jobs than private sector spending—and of these jobs the majority are male-dominated. Whatever growth federal investments produce becomes highly dependent upon the prevailing political winds in Washington.

Of course there are many needs for public sector investments, and important social and sometimes economic reasons for embarking on such investments. Usually we do so with our eyes open about the relative economic inefficiency of such spending. Curiously, though, defense spending has been seen as some sort of exception to this. It has become the sacred cow.

Right now the Pentagon's "investment accounts"—meaning R&D spending, military construction and the procurement of weapons systems—exceed $100 billion per year. That's over $100 billion taken from the private sector and invested in projects and industries favored by military planners. People often have trouble visualizing these enormous figures. Let's put it this way. Say that you drove from Los Angeles to New York and every mile along the way you or the kids threw a thousand dollars out the window. Dealing with singles would be impossible at this rate, but disposing of twenties you'd have at least a second for each one. After a while this might be tedious, but somehow you would keep the bills flying through the purple mountains' majesty, across the amber waves of grain, and so on. Say that about 500 other drivers were willing to help distribute money in the same way. At 10 days for the round trip, all of you would have to drive continuously all year in order to match the Pentagon's current annual investments, quite apart from other items such as military pay and supplies.

No realistic person denies that defense spending is needed; most of us regard a strong defense as imperative. But should we view the cost as a boon to economic development, rather than as a necessary sacrifice to be borne by us all in common? Adam Smith's words still apply today: "The whole army and navy are unproductive labourers. They are the servants of the public, and are maintained by a part of the annual produce of the industry of other people. Their service, how honorable, how useful, or how necessary soever, produces nothing for which an equal quantity of services can afterward be procured."

The issues of the effectiveness and affordability of our current defense policies are somewhat tangential to the main purpose of this book and I will not dwell on them, except to note that a number of serious questions need to be asked in other forums about wisdom and affordability. In what sense is it wise to rely on a hair-trigger system of nuclear weapons with an overkill capacity on both sides? Are we, as former Ambassador to Moscow George Kennan claims "acting like lemmings headed toward the sea"?

With regard to affordability, we have recently learned about the lack of sound business practices that produce light bulbs selling for $44, $600 plastic toilet seats, $7,000 coffee makers, a $9,606 Allen wrench, and a $2,043 nut. Ninety-four percent of military procurement contracts are not subject to competitive bid. Costs are rising so rapidly that conventional forces are actually shrinking, according to some measures of readiness.

My basic economic argument, however, is not that defense spending is inherently unnecessary or bad. I am addressing the frequent claim, or assumption, that military spending is somehow good for the economy. In reality, defense spending is a common sacrifice made for the common defense. Thus, citizens and policy makers must understand how much we are being asked to sacrifice, for what purposes, and what kinds of security these sacrifices will bring us.

The military's accelerating investment accounts magnify the danger of "crowding out" private sector investments by concentrating hundreds of billions of dollars of federal capital in a handful of industries. The raging debate about developing a national industrial policy will soon become moot. With almost no debate or rigorous outside analysis, a national investment policy is being shaped—by the military.

The likely beneficiaries will be defense-related R&D firms plus a few elements of the large-scale manufacturing sector. A negative impact is likeliest to be felt in two areas: non-defense-related R&D which will be at a disadvantage in competing for science and engineering talent, and higher risk investments (such as lending for the start-up and expansion of new enterprises).

Should this occur, we will have succeeded in setting up an industrial policy virtually opposite that of nearly every other Western industrial nation. The tendency toward substantial capital infusions into heavy manufacturing, with reduced capital availability for consumer-oriented R&D and small business, would most resemble the priorities of postwar Britain; the centralized economic decision-making would most resemble the Soviet Union's.

Evidence has mounted in recent years that smaller enterprises and non-defense-related R&D are the engines of growth in mature industrial societies. Smaller firms are now responsible for more than half of all industrial innovations and an enormous share of new jobs on the market. They are also increasingly viewed as the pathway into the economy for new and displaced workers who lack the skills most in demand in prime labor markets.

By contrast, consider the observations of Simon Ramo, a science advisor to President Reagan and a founder of TRW Corporation, in his book *America's Technological Slip.* As Ramo sees it, the spin-off value of military spending is increasingly limited because the U.S. military needs state-of-the-art technological advances applicable to weaponry. Such advances seldom lower costs or increase reliability, which are the objective of much civilian R&D. Thus, for example, the military pours billions into the development of high-speed computer chips—absorbing considerable engineering talent and industrial capacity in the computer industry—while Japan focuses on developing cheaper and more reliable chips with greater storage capacity. More than 70 percent of the latest generation of computer chips sold in the U.S. were built in Japan. Once the U.S. dominated the market.

Surges of heavy government spending in a few industries can also create sharp price increases. In periods of recession, industry is likely to meet the needs of such short-term surges by utilizing obsolete plants and equipment, thus driving up costs through production inefficiencies. Such surges of spending in non-recessionary periods, with less idle capacity, often trigger bottlenecks. First overtime shifts are added, then new capacity is added, meeting the surge in the impacted industries. This not only drives up short-term costs, but leaves more idle capacity after the surge.

In a recent report, Data Resources, Inc. noted that not only does defense spending produce 2,000 fewer jobs per $1 billion spent (compared to $1 billion spent randomly in the economy) but the jobs that are created tend to be highly concentrated in a handful of industrial sectors. The report predicts intense competition between defense suppliers and the rest of the private sector over workers with certain skills, presumably bidding up wages. "Implementation of anything like the Administration's planned expansion of the DoD," the report warned, "will have a major structural impact on the U.S. industrial base and labor markets."

Military spending can slow overall GNP growth. According to the Defense Department, $100 billion spent on defense yields an overall GNP growth of about $220 billion. This "multiplier effect" of 2.2 is exceeded by many other forms of spending. The multiplier effect for spending on exports, for example, is 4.1, according to the Commerce Department. Thus, a dollar spent on exports is about twice as productive overall as a dollar spent on military items.

Sharply increasing defense expenditures also tend to weaken the U.S. balance of trade. An August 1982 report from the Commerce Department concluded that proposed defense spending plans would increase imports and limit exports, "further reducing the U.S. merchandise trade balance." Military goods, especially with respect to their mineral components, are much more import-dependent than most civilian goods. And, because the Pentagon is first in line and can pay top dollar, its orders in industries operating at close to capacity (like microprocessors and industrial ceramics) will displace already backlogged export orders.

The point is that defense spending should not be depicted as a spur to economic development; considerable evidence suggests that there are much more effective ways to spend money, even in the public sector, and most clearly in the private sector. Simply put, an ammunition dump does not generate as many dollars as a machine tool—and it shouldn't be expected to. Defense spending ought to be viewed as an economic sacrifice we make to defend our country, not as a way to stimulate the economy.

I am a Californian. My home state is the third largest arms producer in the world—after the U.S.S.R. and the rest of the United States. My state currently receives over 22 percent of the defense procurement budget and over 40 percent of NASA's contracts.

By almost any measure, California is the "military money capital" of the nation. A National Public Radio report on "Defense Spending" found that one out of five military aircraft is produced in the state; one-quarter of all military communications is built here; one-half of all ordnance and munitions. More than one-tenth of U.S. military personnel is based here.

Weapons manufacture is the largest industry in California—more than twice as large as the farming industry. While total state farm revenues in 1984 reached $14 billion, military contractors received a total of $28.4 billion in prime weapon contracts.

These figures represent the primary influx of capital into the state, as well as hundreds of thousands of jobs. In 1984, almost 700,000 California jobs were directly related to military expenditures. According to a *Los Angeles Times* study, one-eighth of all California workers—and one-third of all scientists and engineers—work directly or indirectly for the defense industry.

Historically, the military budget has been susceptible to "boom-bust" cycles, which are determined by the political mood in Washington and the uncertainties of international events. As a result, defense-related employment in California has fluctuated widely: from 616,000 in 1968 to 440,000 in 1977, and back up to 641,000 in 1982. These employment swings have seriously affected California's defense-dependent communities, such as Los Angeles, Sacramento, Silicon Valley, Oakland and San Diego.

As this process of "defense addiction" accelerates, the state of California may well see its flexibility and responsiveness to market demand seriously impaired by an overdependence on defense-related spending. The ability of industry in our state to respond in a timely manner to rapid changes and developments in the overall economy hinges critically on the private sector's capacity to reallocate resources rapidly. However, if significant portions of that capital are under the direction of decision-makers outside the state—decision-makers who will allocate capital on the basis of non-economic considerations—California's ability to meet its own economic needs and maintain its own economic health will be correspondingly reduced.

Should California begin to develop public policies that are contingent on maintaining or enlarging its share of defense spending, the economic distortions and inefficiencies to which I have referred would probably be worsened. California would be shifting its moorings farther and farther away from the very factors that we know to be associated with sure, healthy sustainable economic growth.

The weapons industry has been vigorously promoted as a stimulus to economic growth, both in California and throughout the rest of the United States. However, this viewpoint is not shared by all California business leaders.

For example, Hewlett-Packard's President John Young has said he fears that U.S. business interests will be "lost in the zeal for defense issues." Steven Jobs, co-founder and former chairman of Apple Computer, Inc. recently stated, "We're making the largest investment of capital that humankind has ever made in weapons over the next five years . . . [This] raises the deficits and, thus, the cost of capital. Meanwhile, the Japanese have shaped their entire society toward raising capital to invest in the next technological frontier—the semiconductor industry. The connections aren't being made in America between things like building weapons and the fact that we might lose our semiconductor industry."

In 1985 the American Electronics Association declared that "we cannot allocate 70 percent of federally funded research and development to defense and severely limit U.S. high-technology exports, even in East-West trade, without paying a very significant price in terms of competitive strength. We cannot siphon off a disproportionate share of our scarce resources to military applications and still stay ahead of Japan in commercial markets."

Representatives of a California economic policy group called Berkeley Roundtable on International Economics (BRIE), which includes business advisory committee members from such companies as Apple, Inc., Intel Corp. and Transamerica Corp., have concluded that the Pentagon can no longer promote commercial competitiveness in innovative industries.

For over a decade, Harvard professor Ezra Vogel, author of *Japan as Number One: Lessons for America*, has warned Americans about the Japanese economic challenge. Obviously, his warning was timely and prophetic. Now, in a sobering sequel, entitled *Comeback*, Vogel paints a chilling picture of a "noncompetitive, troubled" America in 1990, unless the country follows a new "aggressive policy" to restore a weakened economic base.

If California is to play a leading role in the economic future of the Pacific Rim, the state must develop a competitive industrial strategy to supplement the current "weapons-driven industrial policy." Otherwise, California will face increasingly keen competition in computers, communications, satellites, new materials, and biotechnology.

Unless we chart a new course, California and the United States will continue to lose ground and jobs to Japan, which has selected these and other high-tech industries as "targets for the future."

The flood of imports, the loss of jobs and markets for U.S. firms, and the record trade deficits, which have made the U.S. a debtor nation, are cause for national alarm. These economic realities make it clear that an increased commitment to military production has not produced broad-based economic vitality. Unfortunately, the trendline is even less optimistic.

We would do well to heed President Eisenhower's admonition that "there is no way in which a country can satisfy the craving for absolute security, but it can easily bankrupt itself, morally and economically, in attempting to reach that goal through arms alone." Adding that the military establishment is not productive in itself, he pointed out that it "must necessarily feed on the energy, productivity, and brainpower of the country, and if it takes too much, our total strength declines."

Eisenhower cared not only for the security of this country or its economy—as expressed in his slogan of "peace and prosperity"—but also for its creativity. "Today," he once observed, "the solitary inventor, tinkering in his shop, has been overshadowed by task forces of scientists in laboratories and testing fields. In the same fashion, the free university, historically the fountainhead of free ideas and scientific discovery, has experienced a revolution in the conduct of research. Partly because of the huge costs involved, a government contract becomes virtually a substitute for intellectual curiosity. . . . The prospect of domination of the nation's thinkers by federal employment, project allocations and the power of money is ever present and is to be gravely regarded."

These are all among the hidden costs of military spending. To the extent that it is necessary, all of us support it; but let us regard it as a grim need and a sacrifice, not as a wise method for stimulating our economy or for adding to our inventiveness.

Redefining the Role of Business

Ever since World War II, our country's national security has been "managed" by a set of military officers, attorneys, business leaders and others with special concern for international affairs, advised by academic analysts. Within this group there have been some fierce struggles, but for the most part they have taken place within the larger elite consensus. Now some business executives are asking more fundamental questions. It would be hard to persuade an audience that successful capitalists are "soft" on the Soviets. In effect, these critics of the national security elite are asking, "If, as you say, you know what you're doing, how come we're in such a mess?"

Being Dead Is Bad for Business

Lucien Rhodes

DOUGLAS MARSHALL IS THE THIRD GENERATION OF HIS family to own and operate the H. Newton Marshall Co., a union painting contractor specializing in commercial and industrial work. It is a small company, with roughly $1.4 million in annual revenues and an average of twenty-five employees, but it is an important source of his considerable pride. His grandfather started it in 1900; Marshall grew up in the business at his father's side, and finally bought it from him in 1970. He would never do anything that might damage the company's heritage and reputation.

Nevertheless, in April 1983, Marshall wrote a letter to about 100 of his customers, primarily builders and general contractors, which began: "I'm taking a business gamble contacting you about a difficult topic, the nuclear arms race, because I believe its momentum poses an overriding concern." In the paragraphs that followed, he described his personal conviction that the placement of Pershing missiles in Europe would further increase the risk of nuclear war. He then said that he had joined Business Alert to Nuclear War, a group of businesspeople with similar concerns, " . . . because I find that for me the best way to deal with the deep fears generated by this prospect is to try to do something about it." He closed the letter by asking his customers to consider joining the group—which, he said, "has a tremendous potential for influencing public policy, because as businessmen we are perceived as being conservative. Perhaps wishing to arrest the momentum of the arms race and avoid holocaust is the ultimate conservatism."

Doug Marshall is not a man who makes such statements lightly, nor is he some kind of radical zealot who spends his time buttonholing passersby on the street. Most days, in fact, he has all he can do making his rounds of the Boston area bidding for jobs. Still, if he sees an interest or senses an opening, he will talk about peace and about how his attitudes have changed. "For a long time," he says, "I felt that the containment policies backed by nuclear threat had been successful, and I had been content to assume that these policies would work for the next thirty years. But now I think the whole issue has to be rethought. I believe in deterrence, but now we have so much deterrence that the danger of nuclear war—particularly by accident—is increasing, not decreasing."

Marshall came to these convictions by a route that can best be described as circuitous. Twenty-nine years ago, he was an anti-submarine-warfare officer aboard the destroyer U.S.S. *Leonard F. Mason,* having graduated from Harvard University as an ensign in the Navy's Reserve Officer Training Corps program. For most of Marshall's two years in peacetime service, the *Mason* floated tranquilly in the Pacific Ocean, and he peered at the equally placid surface of his sonar screen. But once, in November 1956, there was an ominous stirring in the green, glowing depths below. As Marshall remembers it, the *Mason* was part of a large fleet dispatched to guard Hong Kong against the possibility that the Chinese Communists might grab for it while international attention was focused on the Suez crisis. The *Mason* was three days out of Los Angeles when, at three o'clock in the morning, one of the petty officers under Marshall's command reported the presence of an unidentified submarine apparently trying to penetrate the destroyer screen. Marshall was ordered to arm the ship's depth charges. For about half an hour, the mystery blip drifted along the edges of the sonar screen. Then it disappeared completely.

Following his tour aboard the *Mason*, Marshall went to work in the family business, where he served as a project manager and estimator. Then came the Berlin Wall crisis in 1961, and Marshall was recalled to active duty and assigned to the destroyer U.S.S. *Miller*, which was to patrol the Straits of Denmark between Iceland and Greenland. There the *Miller* would wait for the anticipated deployment of Soviet submarines from Northern Russia into their battle positions in the Atlantic. "That, to me, was the most crucial part of my life because I felt it was not far-fetched to visualize myself as pulling the trigger on World War III," says Marshall. Indeed, he considered the threat so real that he built a small bomb shelter in the basement of his home in Norwell, Mass. But the Berlin Wall incident unraveled into history and the Russian submarines never appeared.

Marshall laughs, but not for long, when it is pointed out to him that twice he has showed up for a war and nobody else came. "It goes to show you how far we've come," he says, "because today that joke's true. The weapons we've got now are so sophisticated that really nobody has to show up. It could start with a bad computer chip. How does it make you feel, trusting your security to the same computers that handle your credit cards?"

During the 1960s, Marshall was a hawk on the Vietnam War—a position that put him at odds with his wife, Jean. "The whole thing is very painful for me," he says. "I was so sure that I was right. My whole training told me that I must be right. But it turned out that actually she was right. I used to think I was an expert, twenty-five years in the Naval Reserve, a lieutenant commander trained in geopolitics. I don't think that way any more; there's another kind of wisdom." As their three children grew older, Jean herself became more and more involved in the peace movement through her local church. Today she is the part-time coordinator of peace-related activities for the United Church of Christ in southeastern Massachusetts. "She's kept me locked into the issues," Marshall says. "I listened to her describe her own work and I kept wondering what I could do."

In the spring of 1981, Marshall and his wife attended a church-sponsored lecture that changed his life. The speaker, a former senior missile adviser to the Air Force, emphasized that the proliferation of nuclear weapons had not increased national security but had, in fact, decreased it. "When he said that," Marshall recalls, "somehow everything congealed for me, and my perspective changed. Until that moment, I thought having more planes, more bombs, more everything was the way to increase security. After the meeting, I sensed I had heard something profound for me, but I couldn't articulate it."

Soon thereafter, Marshall attended a weekend peace conference held at Harvard's John F. Kennedy School of Government. Following the conference, the guests were invited to share their thoughts with one another in groups organized around various professions. Marshall wandered into the room set aside for businesspeople, feeling that here, at least, were people he knew shared a common point of view. "I didn't really know what I was looking for," he says. "I just wanted to get involved in some way. I had really felt safe in exposing my changed views only within my church community. I was afraid of being ridiculed. Seeing those other businesspeople in that room gave me a lot of confidence."

Enough confidence, in fact, to write to his customers. "I took a risk sending out that letter," he says. "Who can tell? Maybe even talking about the danger of nuclear war would be considered unpatriotic in my industry. I didn't want to jeopardize my business, but I felt compelled to speak out on what I regard as the most crucial issue of our time. I was just trying to get something started, a little spontaneous combustion. It's awesome. We go on day after day as if everything's normal. But there's nothing normal about these times."

Sitting in his South Boston warehouse surrounded by hundreds of paint cans, Marshall looks unequal to the enormous threat of a nuclear holocaust. Yet it is precisely in such unlikely champions as Doug Marshall that the future of a new national movement may lie. Until recently, the "little spontaneous combustion" he was looking for simply didn't exist anywhere within the vast and powerful business community on any of the issues considered part of the debate on national security. Other professions spoke up: Physicians for Social Responsibility, the Union of Concerned Scientists, and, more recently, U.S. Lawyers Alliance for Nuclear Arms Control. But business somehow couldn't find its tongue. Only within the past two years has the one segment of society that many observers regard as the most influential been heard from as an identifiable coalition of interests.

At the moment, this nascent movement is little more than a disparate collection of organizations and individuals who, like Marshall, have decided that something needs to be done. The organizations include Business Executives for National Security (BENS), a Washington, D.C.-based trade association; affiliates like Business Executives for Nuclear Arms Control (BENAC), in Philadelphia; and a few independent groups such as the New Forum in Palo Alto, California. (Business Alert, which Marshall joined two years ago, has since merged with BENS.) The individuals involved range from everyday businesspeople like Doug Marshall to full-time entrepreneur-activists like Harold Willens, the former chairman of California's 1982 Bilateral Nuclear Freeze Initiative. All told, the movement probably is made up of no more than 2,000 businesspeople, of which BENS alone accounts for 1,250. But even these numbers, some observers insist, are disproportionately impressive, given business's legendary lack of interest in social issues. "In a society as diverse as ours," says Larry K. Smith, executive director of the Center for Science and International Affairs at Harvard's Kennedy School of Government, "no one segment can do it alone. But the involvement of business in these issues represents an important, essential, and even historic moment."

The conviction that informs this collage of groups and personal initiatives is that many of the techniques used in building a successful business can be applied productively to issues of national security as well. Executives and entrepreneurs, it is said, are practical, action-oriented, and open-minded. They understand the relationship between costs and benefits; they can relate short-term tactics to long-term goals; and they can move quickly from a plan that doesn't work to one that might. "If a guy is hard-headed enough to run a business," says New Forum co-founder Allan M. Brown, also president of Vance M. Brown and Sons Inc., a building contractor in Palo Alto, "he presumably has a rational, pragmatic approach to life, as opposed to being a total visionary or philosopher. I mean, I have to approach everything in a problem-solving way, what's right and what's wrong. Here it's bricks and mortar, and I have to put one on top of the other every day if the business is going to succeed."

Whether business can direct itself to national-security issues is, of course, an open question. Skeptics claim that the business community will never turn out in force, because it is intractably greedy and self-serving. But sympathizers are unperturbed. It has happened before, they say, and it can happen again.

A large part of the impulse powering the current business movement, in fact, traces its source to a surprisingly simple event in the mid-1960s. As Henry E. Niles, former chairman of Baltimore Life Insurance Co., remembers it, his wife, Mary-Cushing Niles, one day called on Joseph D. Tydings, then a U.S. Democratic senator from Maryland, asking him to urge the President to stop sending troops to Vietnam. During the course of their conversation, Tydings said: "I hear from the clergy, I hear from civic leaders, I hear from all kinds of groups. But where are the businessmen?" When Mary-Cushing got home she said to Henry: "Joe Tydings sent you a message. He wants the businessmen to speak out. What are you going to do about it?"

"I thought about it for weeks and weeks," says Niles, who is now eighty-four. "I decided that the thing to do would be to write an open letter to the President and get as many business executives to sign it as possible, because businesspeople were generally thought of as supporting the war. I worked up a draft of the letter and sent it to a few friends I knew in business."

In January 1967, Niles bought space in *The Washington Post* and published the letter, which was signed by 173 business executives. Shortly thereafter, he formed a group out of the signatories and called it Business Executives Move for Vietnam Peace. At the height of its influence, according to Niles, this historic initiative included roughly 4,000 members, and nearly everyone agrees it played a significant role in the movement to end the war. After the war, the organization tried to redefine its goals, even renaming itself Business Executives for New National Priorities. But it was never able to recapture its former influence.

Then, one evening in 1982, Niles had dinner with businessman Stanley Weiss.

For most of his life, Stanley Weiss had thought little about war, peace, or anything else having to do with national security. He was too busy living a fairy tale. In his early twenties, Weiss saw the movie *The Treasure of the Sierra Madre,* in which Humphrey Bogart searches for gold in Mexico. Weiss liked the idea so much that he went to Mexico to relive Bogart's adventure. For three years, from 1951 to 1954, he prospected for gold in the Central Plains of Mexico near San Luis Potosi. He never found any. But he learned his rocks, and he learned to survive. "They were dangerous times," he says. "Everybody was always armed. You know, I tell people now that nuclear war is bad for business because being dead is bad for business. But that's where I got that line. I mean, if I'd gotten shot, I never would've had a business."

In 1954, a cab driver in Charcas, Mexico, told Weiss that he knew of a vast mound of manganese that he would show Weiss if they could form a partnership. Weiss agreed immediately: he knew that in many ways manganese—which is essential, for example, in the production of steel—was as good as gold. After a two-hour march along a dirt trail, Weiss saw what he describes as a "mountain of manganese," and, in fact, the mine he soon opened bore that name. Borrowing $5,000 to get started, he began shipping tons of the stuff to processors. "First I was starving to death," Weiss says, "and then I struck it rich." He was not yet thirty.

In the years that followed, Weiss opened his own processing plant in El Paso, and incorporated his efforts as American Minerals Inc., of which he is still chairman. But manganese gave Weiss more than an exotic tale to tell and more than an economic power base; it also set in motion a series of interrelated events and reflections that ultimately led him to Henry Niles. The linkage began to build in 1975, when Weiss wrote a definitive text on manganese. To his surprise, he was immediately recognized as an expert on strategic minerals, and was subsequently invited to spend a year as a fellow at Harvard's Center for International Affairs. There he began to see that if manganese was important to the nation's security, then the very idea of security must include much more than its common definition as military might. He also began to think he might have a special role to play in getting the business community involved in national-security issues.

Weiss spent the next year searching for groups or individuals who might already be working on programs involving businesspeople. It was a sparse landscape. "I kept thinking there must be somebody who knew more about these things than I did," Weiss recalls, "but I never found anyone. There were people who said they'd help, but they never did." Then he thought of Henry Niles. Weiss had heard so many garbled and contentious impressions of what Business Executives for New National Priorities was or wasn't still doing that he decided to ask the man himself. In the spring of 1982, the two men had dinner together at the Cosmos Club in Washington, D.C. There, Niles asked Weiss to take over what remained of the group. Weiss refused, but said he would consider starting a new organization.

Among other things, Weiss wanted the new group to be distinctly nonpartisan, unlike Nile's group, which he felt was viewed as a "left liberal" encampment. He also felt that the group should be a fully accredited trade organization, because under the Internal Revenue Service's rules, that was the only way the group could accept tax-deductible contributions from members, yet preserve its status as a political lobby. In addition, as a trade association, the new group could work more easily with established trade organizations.

During the summer and fall of 1982, Weiss refined his plans, with substantial help from a few early supporters. Niles liked the plan so much that he gave his mailing list to Weiss and discontinued his own group. From his Washington apartment, Weiss sent out a mailing to Niles's list of roughly 2,000 names announcing the formation of Business Executives for National Security and soliciting memberships. "We had a fair response," Weiss says. "Unfortunately, about fifty percent of the people on the list were dead and maybe another twenty-five had moved. If you say we got about 200 members that would be about right."

Today Weiss enjoys pointing out that by granting BENS trade-association status in 1983, the IRS by definition agreed that the organization was working for the basic business interests of its members. "In other words," Weiss says, "they agree that being dead is bad for business."

While Weiss was licking stamps, Harold Willens was completing his role as chairman of the successful bilateral nuclear freeze initiative, which called on the United States and the Soviet Union to agree to an immediate and verifiable halt on the testing, production, and deployment of all nuclear weapons, missiles, and delivery systems. For Willens, it was an important milestone in a résumé of personal political activism that had already spanned twenty years.

Willens is a campaign unto himself. He has a store of personal experience and business accomplishments that, in many ways, make him the ideal protagonist in the unfolding drama of a new national movement. He was born in Russia in 1914, during the tumult of the Russian Revolution, in the village of Chernigov. In his earliest memories, he is hiding under a bed as marauding soldiers burst into his parents' house demanding food and valuables at swordpoint. Fleeing Russia, his family emigrated to the United States, settling first in a mixed ethnic ghetto in the Bronx. Their life there, he recalls, was one of unrelenting poverty and fear: Willens, who is short and slender, got "beat up every day."

When he was thirteen, Willens' family moved to another ethnic ghetto, Boyle Heights in east Los Angeles. Life was still tough, but gradually Willens began to accumulate some pleasant memories. Three years after he graduated from high school, he bought a small retail route for $250, which consisted of a truck and a list of customers to whom he sold mayonnaise, pickles, tamales, and a variety of specialty foods. He was a good salesman, and soon bought larger trucks and larger routes. He got married, bought a small house, and started a family. He even went back to school, graduating Phi Beta Kappa from the University of California at Los Angeles in 1939.

During World War II, Willens served as a Japanese-language specialist in the U.S. Marine Corps. Part of the early American occupation force in both Nagasaki and Hiroshima, he sifted through his fingers the fine dust that had once been buildings and people. He wrote his impressions in long letters to his wife. But as the war came to an end and he began his life again, even the memory of the only use of nuclear weapons against other human beings faded.

Willens had sold his house and business before he went off to war, using the money to buy two neighborhood grocery stores. If he was killed, he reasoned, his family could still live comfortably on the rent money from the two stores. The stores were located on a relatively quiet section of Wilshire Boulevard in Santa Monica. After the war, the boiling commercial activity spilling out of Los Angeles engulfed the property, and Willens was able to parlay his small patch of real estate into a kingdom that now includes, among other holdings, entire blocks of Wilshire Boulevard and several shopping centers. Only ten years out of the service, he found himself a multimillionaire. "As the saying goes," he writes, "now that I had done well, I wanted to do good."

Like Niles, Willens found his attention drawn to the escalating war in Vietnam. Not only did it represent his first major engagement as a political activist, it also introduced him to the same special constituency that Weiss found: businesspeople. As Willens remembers it, he first became aware that business had a particularly important role to play through a suggestion from Marriner S. Eccles, who had been the first chairman of the Federal Reserve Board, serving from 1936 until 1948. Willens had called on Eccles to ask for a donation to the Center for the Study of Democratic Institutions, a Santa Barbara think tank with which he was associated. During the course of their conversation, he told Eccles what he had learned about the war in Vietnam. Eccles thought the information should be distributed more broadly, and suggested that Willens could be particularly effective in reaching other business leaders.

The notion made sense. "If you come up before a business audience sounding like some kind of hippie," Willens says, "they'll write you off. But you see, I look like them, I'm a multimillionaire, I went from poverty to affluence, and they respect that." After his meeting with Eccles, Willens met Henry Niles, and together they created Business Executives Move for Vietnam Peace, an association that for Willens combined the right issue with the right audience. He became chairman and ranged the country giving speeches and raising money.

When the war in Vietnam ended, Willens kept active: he involved himself in Presidential campaigns; founded or helped to found such organizations as the Businessman's Educational Fund, the Center for Defense Information, and the Interfaith Center to Reverse the Arms

Race; served as a delegate to the United Nations Special Session on Disarmament in 1978; and, of course, served as chairman of the bilateral nuclear freeze initiative in California. While working on the initiative, he read the results of a survey commissioned by *The Wall Street Journal* and published in September 1982. The survey asked executives: "Do you favor a unilateral freeze by the United States on the production and deployment of nuclear weapons?" The survey still irks Willens, who thinks the newspaper was engaging in some kind of "dirty pool" meant to discredit the California initiative. The issue of the moment was a bilateral, not a unilateral, nuclear freeze, and an immediate unilateral nuclear freeze was not then, and is not now, part of mainstream thinking. Nevertheless, although an overwhelming number of the respondents opposed such a freeze, Willens was astonished to find that 36 per cent of the executives in the smaller companies surveyed favored it, as did 27 per cent of the medium-size-company executives and 14 per cent of those from larger companies. Given that much support for such a radical approach, Willens reasoned, a more realistic and moderate program could probably convince a telling majority of the business community.

If the business community could be convinced, moreover, it would be disproportionately influential. When he was speaking out on Vietnam, Willens often found that he got more attention from the press than he did from his own constituency. "I learned quickly that to be a business executive and to oppose military spending and military involvement was like being a two-headed calf," he says. That, in turn, led Willens to a new view of the importance of business. "It became clear to me that businesspeople acting as citizens could get tremendous attention on a controversial issue—and could become a 'trimtab.'"

Willens borrowed the trimtab concept from R. Buckminster Fuller; it refers to a small amount of leverage, precisely placed, that ultimately produces enormous effect. On ocean liners, for example, the trimtab is a small rudder attached to the huge main rudder, which, because of the mass and momentum of the ship, is difficult to turn. The trimtab, however, responds quickly to even a small influence and then turns the main rudder, which in turn alters the course of the entire ship. Willens set out to write a book, which he called *The Trimtab Factor*. The cover blurb reads: "Our ship of state is taking us into very dangerous waters. Business leadership can be the trimtab that will change our direction before it is too late."

The most challenging part of Willens' book is his presentation of what he calls an "incremental nuclear weapons freeze." He outlines a series of steps by which the United States can take the initiative in proposing moratoria—first on the testing of nuclear weapons, then on the flight-testing of nuclear-weapons delivery systems, then on the deployment of any new nuclear weapons systems. And he concludes with a proposal to reduce existing nuclear arsenals drastically. If the Soviet Union doesn't respond to the U.S. initiative within a reasonable period of time, in Willens' plan, the United States would no longer be bound by its proposal. Willens believes this program is particularly well-suited to the inherently conservative disposition of business: it represents a more moderate adaptation of the 1982 Bilateral Nuclear Freeze Initiative, which has been endorsed by more than twelve million voters in nine states. It is also, he says, politically achievable. Moreover, it "represents a methodical process for breaking the momentum of the arms race with no real risk to our national security."

Since February 1984, Willens has turned into a sort of one-man movement, promoting his "primer" with a double-barreled marketing strategy that relies partly on the efforts of William Morrow & Co., his publisher, and partly on the "financial participation of friends who believe in what I'm doing." Immediately after publication of the book, he galloped through the usual promotional tour, overlaying it with additional visits, which he engineered personally, to newspapers, magazines, and business groups. He also put $60,000 of his own money into the marketing effort, and took personal responsibility for distributing 22,000 copies of the book. He had copies hand-delivered to every member of Congress, accompanied by a flattering letter of introduction signed by Senators Mark Hatfield (R-Ore.) and Edward Kennedy (D-Mass.). He has since sent copies to most college and university presidents, to religious leaders, to presidents of Chambers of Commerce, and to various media executives, editors, and journalists. Some executives have been so impressed by Willens' book that they have also taken to distributing books at their own expense. James Jensen, president of Seattle-based Thousand Trails Inc., bought 1,000 copies. And Lawrence Phillips, president of Phillips-Van Heusen Corp., is, according to Willens, sending books to the chief executive officers of the *Fortune* 1,000 companies. Phillips was among a list of twenty-five executives who, along with Willens, recently advertised the book in a two-page spread in *The New York Times* at a cost of $60,000.

Willens says he will feel fulfilled if, within the next twelve months, his initiative inspires the formation of twenty-five regional groups such as Allan Brown's New Forum, which he considers a fine example of an influential grass-roots movement. Brown, a soft-spoken, self-effacing man, was flattered when he learned of Willens' opinion, but disclaims any such cosmic ambitions. He agrees, however, that being able to sign letters as the president of a fifty-seven-year-old general contracting company with revenues of $25 million to $35 million a year does give him a little extra clout. "People are admired because they have climbed to the heads of corporations," he says. "They are believable because they have obviously been successful in making decisions—and military matters are also practical decisions."

The New Forum, which began as no more than a vague impulse "to do something tangible," does exemplify business involvement in its most unpretentious and elemental form. In the fall of 1983, Brown and several friends invited four like-minded people to a breakfast to explore how they could educate people about the danger of nuclear war. Friends invited more friends from business and other professions to more breakfast meetings, and, after several pounds of coffee and rolls, the New Forum emerged, replete with a newsletter, study groups, and a schedule of monthly meetings. Brown says the group, which now has more than 200 members, is currently considering various ways of amplifying its concern with direct political action.

BENS, meanwhile, has emerged as a dependably generic representation of the business movement at the national level. Its membership list reflects support from a broad cross-section of American business: Sidney Harman, chairman of Harman International Industries Inc. and former Undersecretary of Commerce; J. Richard Munro, president and chief executive officer of Time Inc.; real estate developer

James W. Rouse, chairman of The Rouse Co.; John C. Haas, vice-chairman of Rohm & Haas Co.; Mortimer B. Zuckerman, Boston real estate developer and owner of *The Atlantic* magazine; Robert Stuart, chairman of National Can Corp.; and, since late 1983 when Business Alert merged into BENS, Douglas Marshall, president of H. Newton Marshall. "We've been very careful to reflect American business thinking in all that we do," claims Stanley Weiss, "particularly the entrepreneurial part of it. It's part of the entrepreneur's very nature to take a new idea and make it succeed."

There are three ideas that BENS would like to see succeed: reducing the danger of nuclear war; promoting a "strong, effective, and affordable defense"; and working for a more productive relationship with the Soviet Union. Weiss felt from the start that although the implications of a nuclear war must transcend all other issues, it wouldn't make good business sense to concentrate on nuclear arms control exclusively. For one thing, nuclear weapons account for only 15 to 20 per cent of all weapons expenditures. For another, he felt, focusing on nuclear arms control alone would obscure the importance of a greatly expanded definition of national security. "National security is always thought of in one-dimensional terms, as military might," Weiss says, "when it really rests on three legs: military, economic, and the condition of the country's physical and human infrastructure." Fat military budgets, financed as they are at the expense of private sector investment funds, actually hurt business, he believes; American corporations have trouble competing with their counterparts in Japan and West Germany precisely because those countries have military budgets much smaller than our own, and can thus invest heavily in the private sector.

So far, BENS has gotten itself involved in most of the activities one ordinarily associates with a trade association. It publishes a newsletter, *Trend Line,* which informs members about relevant news on Capitol Hill, and it sends out periodic "Action Alerts" on important congressional votes. Its educational fund conducts studies of such issues as the relationship between defense spending and the economy, and it runs a speakers' bureau. As to whether it will win a more widespread following, either among members of the business community or among

Capitol Hill policymakers, most observers agree it is too early to tell. But BENS has already secured a reputation as an important resource for representatives, senators, and trade associations looking for factual and intellectual support as they frame their positions on defense policies. John Motley, director of federal legislation for the 570,000-member National Federation of Independent Business, who worked with BENS in fashioning his association's position on the Bipartisan Budget Freeze Proposal, says BENS "provided good information and valuable arguments which could be used to convince senators that a one-year freeze on defense spending wouldn't be a disaster."

Weiss, for his part, is modest about the organization's accomplishments. "We're not exactly a household name," he says. "But then again, we're barely two years old."

"Like every important social and political change in this country," Harold Willens says, "the desire to end the arms race must start at the bottom. I always tell people: 'If the people lead, in time, the leaders will follow.' This is what has to happen on the nuclear arms race issue so we can end this insanity before it puts an end to us." But whether business, which Willens sees as the trimtab of America, cares enough to shoulder its presumed responsibilities is open to question. Rear Admiral Gene LaRocque, USN Ret., a founder and now director of the Washington, D.C.-based Center for Defense Information, goes so far as to suggest that if business doesn't break its silence, it will be guilty of a grave sin of omission. "Businessmen alone can't do it," he says, "but in my view businessmen are perhaps the most important group, if they get their act organized better than they have. Businessmen are very slow in coordinating on this issue. If the business level of interest stays the way it is today, they will—also in my view—be largely responsible for our failure to move away from this arms race, because they are potentially the most influential."

Meanwhile, in the basement of his house in Norwell, in the small bomb shelter he built after being recalled to active duty in 1961, Douglas Marshall also wonders if anybody out there is listening. Marshall rummages through the shelter's sparse contents, annotating them with a certain bemused nostalgia. There is the chemical toilet. A sheet of plywood lies on the struts that once supported the bunk beds. The air-intake pump still works. There is a bookshelf in one corner; next to it is a short two-by-four that would have slipped between two steel hooks and closed the door against intruders. A child's drawing easel is leaning against the wall. On the top of the easel, one of Doug's children, now grown, had once written the word "peace" in yellow paint.

"It's been kind of a storeroom for a long time now," he says, closing the door. "Only good for growing mushrooms really." Then, as an unexpected afterthought recalling an earlier conversation, he says: "You know, only one person ever did write back to me about that letter. No one else even mentioned it."

4 Shifting to a Global Peacetime Economy

The world is now divided into two main trading blocks. In a truly global economic order, goods and services would be exchanged freely across *all* borders, without reference to ideology. At present, the two superpowers exchange very little, apart from grain transactions. Should the U.S. attempt to expand its trade with the Soviet Union substantially?

In the U.S., one group says no: If the Soviets want something enough to buy it with hard currency, we can impede them by refusing to provide it. Otherwise, this group says, we'd be "selling them rope with which to hang us." Many types of high technology that we take for granted might be swept along in the exchange, allowing the Soviets immediately to increase the effectiveness of the military forces they already have. Even apart from items of high technology, the Soviets would probably apply any economic gains to their traditional goals of military power and subversion of other regimes. And even if the Soviets appeared peaceful for a while, they could later use their increased economic power for aggressive ends.

In this view, the self-constricting nature of the Soviet economic system is a major advantage to the West. It's not that they lack resources—while we're importing a third of our oil, for example, they've been exporting fuels. It's not that they lack good scientists or engineers. The trouble is that they lack the discipline and flexibility of the market, and the scope that our system offers for individual initiative. Why let them buy from us what their stultifying system makes it hard for them to develop at home? Besides, their system is very hard to do business with. And trade has never prevented wars—for example, the European powers were all trading in 1914.

The argument against increasing trade continues by noting that the Soviets, facing a labor shortage, desperately need our technology to increase output per worker. Why help them out of this demographic trap? As the Soviets are forced to continue restricting consumer goods, their own population will grow even more sullen and, despite Gorbachev's campaign against vodka, they will drink more instead of working harder. A faltering economy will further limit the appeal of Marxism to Third World countries.

In any case, some say, the Soviets need our technology more than we need their money, and we should agree to trade only in return for a major "human rights" concession, in particular the right of more Soviet Jews to emigrate.

Also, say opponents of increasing East-West commerce, if the U.S. undertook substantial trade with the Soviet Union, Moscow could later pressure us by threatening disruptions in it, while the U.S. government, in seeking to retaliate, would be lobbied by all the economic interests who profit by trading with the Soviets. Just as Midwestern grain farmers and dealers already oppose embargoes by the U.S., so a much broader economic constituency would campaign against any future attempt to pressure the Soviet Union by reducing our exports.

Instead of offering to provide the Soviets what their economy lacks, this argument concludes, we should deprive them of everything except humanitarian trade, such as the selling of grain. Some strategists with influence in the current Administration add that we should step up the arms race to keep the Soviet economy under maximum stress, if possible to break it.

On the other side of the debate there's a moderate argument and there's a strong one. The moderate version says that most of what the Soviets can't get from us, they buy from the Japanese and Europeans anyway; that strategic items, once defined, can easily be excluded from trade with the Soviets; that we would profit by regularizing our economic relations with the other superpower; that the Soviets are already able to do to us militarily most of what they would be able to do if they were richer; that their need to compete in selling goods in the West might encourage changes that would undercut their ideological appeal on the Third World, such as it is; that our best hope for peace lies in encouraging and assisting the Soviets to shift from a quasi-war economy to a consumer economy; and that the more our goods become valuable to the Soviets, the more careful they will be about disrupting the larger relationship.

In what I'm calling the strong version of the pro-trade argument, proponents ask, "Why not offer to increase trade sharply while also agreeing on deep cuts in arms?" Under these conditions, the Soviets could satisfy their consumers to a degree never dreamed of (except in those recurrent Soviet fantasies known as Five Year Plans). At the same time, they would increase their security to the extent that arms reductions were arranged so as to decrease the fear of surprise attack or "crisis instability." It's even possible that economic adjustments required by reducing military budgets could be eased through trade between the superpowers.

If trade with the West allowed a rise in the living standard of Soviet workers, the beneficiaries would have an interest in further trading. While it's true that the Soviet regime is not representative, except on paper, workers can "vote," in a sense, by adjusting their degree of commitment to the job. (Anyone who reads the barrage of exhortations in *Pravda* will not suppose this feedback is going very well for the regime.) And if Western consumer goods were available (beyond Pepsi), Soviet workers could also vote with their rubles.

A change of this magnitude may seem incredible, but think of old enemies such as France and Germany joining in the Common Market, and think of the dramatic opening between the U.S. and China.

In 1951, sociologist David Riesman published "The Nylon War," an essay that brilliantly combined utopian vision with satire about superpower relations.[1] Taking the form of a brief history of a daring operation conducted with CIA funds, this fictional account starts with "an idea of disarming simplicity: that if allowed to sample the riches of America, the Russian people would not long tolerate masters who gave them tanks and spies instead of vacuum cleaners and beauty parlors"—or as we might add today, compact disk players, home computers, and microwave ovens. If the U.S. could deliver samples to Soviet consumers, their rulers "would be forced to turn out consumers' goods, or face mass discontent on an increasing scale."

How to bring this about? One could not send out mail order catalogues, after all. The masterminds of the Nylon War decide to parachute the goods into Russia from cargo planes. As the first stoves, wrist watches, and toys are delivered over Soviet cities, the regime falls all over itself in responding. It declares that the attack never occurred; that it was a provocation to goad the Soviets into war; that the gifts show the American system, unable to sell its output, is near collapse; that the goods were actually made in Russia; and that they are poisoned by the Americans. Meanwhile, of course, Soviet consumers gather up the goods, trade them on the black market, and begin to ask, "Why don't our leaders invest more in consumer goods instead of weapons?"

Without recounting the rest of Riesman's very funny, thought-provoking story, I want to note that its central point has become even more relevant thirty-five years after it was written. So far we have been competing with the Soviets mainly in the one area where their economy works—the design and production of weapons in great quantity. Why not attempt to shift the rivalry to areas in which we hold a commanding lead—consumer goods and other non-military equipment—and in which the Soviet Union can supply things we need. Imagine a rivalry that was economic rather than military, based upon fair exchanges instead of dangerous threats.

Some of the objections to increasing trade are taken up in the following chapters; in any case, a brief introduction is not the place to debate the entire issue. The central question is whether the Soviet Union can be coaxed, through trade among many other means, toward a less volatile, more constructive relationship with the West. Can each side stop threatening the other and instead share its comparative economic advantages?

What's needed is a shift from what political scientist Richard Rosecrance calls a "territorial" system to a trading system. Instead of constantly warning the other, "We can destroy you," we would keep saying, as our primary message, "Your side and ours could help one another in many ways." In *The Rise of the Trading State,* Rosecrance writes:

"Since 1945, the world has been poised between two fundamentally different modes of organizing international relations: a territorial system which harkens back to the world of Louis XIV and which is presided over by the USSR and to some extent the United States, and an oceanic or trading system that is the legacy of British policy of the 1850s and which is today organized around the Atlantic and Pacific basins."[2]

Basically, territorial states seek to encompass the resources, raw materials, and markets necessary to be self-sufficient. In contrast, trading states depend upon their inventiveness, ingenuity, work discipline, and sensitivity to the market, as they seek the rewards of comparative economic advantage. "Assuming that trade is relatively free and open, they do not need to conquer new territory to develop their economies and to provide the essentials of a consumer existence." Leading examples are Japan and Western Europe. "Today West Germany and Japan use international trade to acquire the very raw materials and oil that they aimed to conquer by military force in the 1930s."

What if the Soviet Union could share similar economic benefits while satisfying its legitimate security needs through a global agreement? If it's fated to be an empire, why not an empire of trade? In terms of average income, how many Soviet citizens would not trade economic places with the Japanese? Yet the Japanese have very little oil or other domestic energy sources, little arable land per capita, few of the resources plentiful in the Soviet Union, and no armed forces beyond home defense units. Can the territorial empires follow the Japanese example and learn to compete as inventors, producers, and salesmen instead of as bullies?

Questions such as these lead to many others, only a few of which will be touched upon in this section. However, the following chapters at least sketch an alternative to our present system of trying to keep the Soviet Union economically retarded by denying them technology and by making sure they will squander a large fraction of their gross national product on the arms race.

—Craig Comstock

1. David Riesman, "The Nylon War," reprinted in *Abundance for What? and Other Essays* (Doubleday, 1964).
2. Richard Rosecrance, *The Rise of the Trading State: Commerce and Conquest in the Modern World* (Basic Books, 1986), page 16.

In *The Making of the President 1960*, Theodore White drew attention away from the outer spectacle of politics, to the organization, tactics, and deals that lie behind it. The success of his book encouraged many of his colleagues in the media to report less on rhetoric and more on backroom politicking that often contradicted the rhetoric or was largely detached from it. Trying to follow White's lead, some reporters have been neglecting not only rhetoric but the ideas and presuppositions that affect what politicians do and how they vote. Yet as John Maynard Keynes reminded us, statesmen often merely act out the theories of some forgotten scribbler.

When reporters ignore ideas, they are reduced to describing current reality, at the expense of examining various alternatives to it. It amounts to a censorship directed against the imagination. One exception to this blackout is ideological magazines; another is some of the journalism of the "counter-culture." William Greider writes for *Rolling Stone* which, along with interpretations of rock lyrics, gossip on the career crises of young performers, and news on the rise of compact discs, also runs sassy, breaktime essays on politics. In January 1985, it published the following report on ideas from the World Policy Institute for an alternative approach to security.

Our purpose in reprinting Greider's report is not to endorse any particular approach to security, but to illustrate what an alternative looks like. Apart from foreign policy specialists, most Americans have seldom or never been introduced to a wide-ranging alternative.

The Economics of Security

William Greider

AMERICANS ARE SO BRAINWASHED WITH FEAR OF RUSSIA that it might seem futile in the present climate to focus on rational alternatives to the Cold War. Yet earnest, able people are doing that, trying to construct a new foreign policy that matches reality. One challenging set of answers—at least a first step toward reason—is being published by a small New York foundation, the World Policy Institute, in its new quarterly, *The World Policy Journal,* and in an ongoing blueprint for new policies called the Security Project. "Our job,"

Archibald L. Gillies, the institute's president, explained to me, "is to show that we can have a different set of policies that hang together and really address world realities, that don't give away the store and that will work better than the national-security policies we have now."

In the Security Project, Gillies and his colleagues propose three fundamental perspectives that ought to displace America's obsessive fear of the Russians:

First, the economic forces that threaten American prosperity have to be viewed as complex global problems in which everything is interconnected, from the horrendous federal deficits to the loss of American industrial jobs, from the Third World debt that threatens the U.S. banking system to our bloated defense budgets, from oil shortages to hunger. Because the world is now one marketplace, both for products and for labor, we can't solve the problems of Detroit without also dealing with the problems of Brazil. This is not sappy do-good sentiment—it's calculated self-interest. Brazil's middle class buys the durable goods that America makes; if Brazil is going down the tubes, so are a lot of American jobs. Our economy will grow faster if theirs does.

Second, the U.S. has to stabilize its relations with the Soviets, offering both sides the immediate economic reward of shrinking defense budgets and new trade. The Russians need what we have to sell, mainly modern high tech, and if we don't sell it to them, Japan and Western Europe will.

Third, America needs to adopt a noninterventionist policy toward Third World countries. Of course, if you believe that every revolutionary struggle or change of governments is a Commie plot run by Moscow and designed ultimately to threaten us, then the idea of reducing our military strength worldwide sounds like lunacy. But the realities of Vietnam, Iran, Nicaragua, Chile and the Philippines, to name a few, do not match that simple-minded fear. Our national security would not be affected in the least if, for instance, democracy were restored to Chile or even if a Marxist government took over. As of now, we have about 550,000 troops overseas, many of them stationed in tripwire positions like the one in South Korea, where Americans would not really wish to fight a war if one started. This is dangerous. It also puts us in permanent conflict with the natural aspirations of impoverished peoples for economic and political justice.

Gillies is surprisingly optimistic about selling these ideas to Americans, notwithstanding the recent election returns, because he and his colleagues see the competing economic and political realities closing in on the Cold War dogma. The danger, he concedes, is that Americans might pull back from the world into a "Fortress America" mentality that would be repressive and self-destructive.

The institute's Security Project has outlined what is actually a modest proposal for a gradual reduction of the American military—changes like shrinking seventeen army divisions down to eleven, reducing strategic nukes to about 6,000, settling for six aircraft carriers instead of fourteen. One of the designers, a defense expert named Gordon Adams from the Center on Budget and Policy Priorities, argues that in this cutback the Pentagon would not lose any of its basic capabilities: deterring Soviet nuclear attack, keeping open the world's sea lanes, defending the continental United States, even intervening militarily with the marines. But the savings would be extraordinary—as much as $470 billion over the next five years, plus another $1 trillion accumulated in savings during the following five years.

"Just imagine," Gillies said, "what would happen if we had $1.5 trillion to use in the next ten years on all the other problems that face us: deficit reduction, refinancing world debt, rebuilding our basic industries and cities, financing high-tech industries, retraining workers. It would change the whole nature of the debate on every other problem."

If these dollar savings sound like pie in the sky, they are really quite conservative estimates. The defense budget proposed by the Security Project follows the same projections for military spending that the government itself was using in the Nixon-Ford administration before the Cold War was reborn in the last years of the Carter presidency. The money aside, Gillies makes this crucial point: "If we take these steps, our security will increase, even vis-à-vis the Soviets. That's the starting point. It's in our interest. We'll be safer."

The most dramatic example of how post-Cold War thinking would increase our national security, even as it improved our economic prospects, is in Europe. Americans seem unaware of the fact that at least $133 billion of our annual defense budget goes to defending Europe—picking up the tab for Germany, France, Britain and the other countries even as we compete with them for world trade. There is something wrong in that equation, but a decade of efforts by American presidents to persuade our NATO allies to spend more on defense has produced almost no change. They know a good thing when they see it.

In military terms, the Reagan administration keeps adding to both the nuclear and conventional forces positioned in Europe, mostly in Germany, as though we were still supposed to believe that the Red Army is poised to sweep across the German plain and drive for the English Channel. Almost nobody in Europe takes such a scenario seriously any longer, even among defense strategists. Indeed, the U.S. military presence is as endangering as the Soviet threat it is designed to counter. You can't fire off many tactical nukes without obliterating West Germany, not to mention our ten divisions stationed there.

Sherle R. Schwenninger, editor of *The World Policy Journal*, offers an alternative: "If we define the problem in central Europe not as a Soviet attack across the German plain but as the potential for conflict by accident as Soviet power erodes in Eastern Europe, you don't want a military posture that will seem like a provocation. You want a defensive posture that is serious enough to be convincing—meaning it would incur a very heavy cost to the invader—but does not add to the military tensions in Eastern Europe. If the Soviets have more normal relations with the U.S., the pressures for change in Eastern Europe are less likely to produce Soviet paranoia about its legitimacy and global image as a superpower."

In tangible terms, this means pulling troops and weapons back from the front line in central Europe, "the fault line of potential global conflict," as Schwenninger calls it. The idea in different forms is seriously discussed in Europe, if not in America, not just by peacenik orators but by governments. In the Balkans, the governments of Albania, Yugoslavia, Romania and Greece (itself a member of NATO) have proposed that the two superpowers honor their countries as a nuclear-free zone where no nukes could be stationed. In the north, Sweden, Finland and Norway are discussing a similar posture. Even in central Europe, a less dramatic proposal is gaining support—the creation of a nuclear-free corridor, 100 miles wide, across Germany.

The economic benefits for America are obvious: a huge easing of our defense burden, presumably phased in over many years, as we gradually bring home the ten divisions now permanently deployed there (the Security Project envisions withdrawing only two of those divisions in the next five years). This would be done, as Gillies argues, step by step, watching both the Soviet response and the concerns of Western European politicians as they at last take responsibility for their own national defense.

America also needs to revive the promising economic initiatives that Richard Nixon undertook a decade ago when he encouraged U.S. trade with the Soviets. Because of its population decline, the Soviet Union expects labor shortages in the future, which means it is desperate to rebuild its industrial facilities with high-tech automation. Our allies in Japan and Western Europe are already developing this vast market; they're delighted that Cold War hysteria in the U.S. keeps American businesses from competing with them.

The common-sense solution to this situation—American business plunging full steam into the Soviet marketplace—may not be as remote as it sounds. The Commerce Department recently reopened trade talks with the Soviets, focusing on some of the same projects abandoned in the mid-1970s when détente collapsed. American companies have the best technology in the world for mineral extraction and production—coal, oil, natural gas—and the Russians want to buy it.

Right-wingers in Washington are already denouncing this initiative as the work of "tradeniks." These Cold Warriors would like to cut off the Commies completely, even refusing to sell them American grain. But if one thinks about it, it is in our self-interest to help Russia develop its resources. For the next twenty-five years or so, the threat of energy shortages will continue to hang over the world's prosperity, including ours, at least until alternative technologies are developed and operating. The Soviet Union has the world's largest untapped oil fields, which, when developed for greater production, are bound to reduce the pressures on world oil markets. As our own domestic sources decline, it makes sense for us to burn Russian oil first and save our own precious reserves.

American self-interest in the Third World is more complicated, but it follows from the same clear-eyed pragmatism: what is good for them may also be good for us. In case after case, if we set aside the overheated fears of Soviet empire building, Americans would discover that our fortunes are closely bound to the future of those struggling nations. Today, we depend on them less as sources of raw materials and more as markets that will buy our exports.

Yet the dominant Cold War strategy ignores the economic issues and repeatedly puts us at odds with Third World political aspirations, whether reformist or revolutionary. Reagan's UN ambassador, Jeane Kirkpatrick, once argued that if the U.S. did not support dictators like Somoza in Nicaragua, "our enemies will have observed that American support provides no security against the forward march of history." Why *should* Americans want to thwart the "forward march of history," anyway, since we are a forward-looking people ourselves? If we were less insecure about the Soviet Union and more confident of the benefits of democratic capitalism, we would know that history is on our side.

Jerry W. Sanders, a senior fellow at the World Policy Institute, has called this crusade to contain Soviet influence around the globe a form of isolationism, "ideopolitical isolationism." The phrase is awkward, but if you glance around the world, it sounds right. The Cold War theology compels the U.S. government to make common cause with the apartheid government of South Africa or the brutal repression of Pinochet in Chile or the murderous regime of Marcos in the Philippines. In each case, we are indeed isolating ourselves, not just from the future generations of those countries, but from our moral values. Common sense argues that we should normalize relations with smaller nations like Vietnam and Cuba, not to embrace their ideologies but simply to open an adult dialogue with them. They need us—the benefits of trading in our markets—more than we need them.

The crucial point about these economic problems is that in the past Americans have not thought about them as foreign-policy issues. Yet finding solutions to these problems will have more impact on our real security than all of the tiresome debates over how many new weapons to buy. That, perhaps, is how our fear of the Soviet Union hurts us most of all—it spooks an entire nation into ignoring its real perils.

This set of Cold War alternatives has, of course, one major drawback. None of the above ideas has a prayer of being seriously considered in the Washington of today. This year, Congress will doubtless trim a few billion from Weinberger's defense budget, and, if people are not deceived by Reagan's sudden enthusiasm for arms control, Congress may even kill the wasteful MX missile system. Those worthy fights will go on in the trenches, but none will challenge the deeper premises of the Cold War or embrace the solutions raised by new thinkers like the folks at the World Policy Institute.

In the 1984 campaign, we got some glimpses of original thinking in Senator Gary Hart's opposition to military intervention in the Middle East and Central America. We saw a clearer outline in many of the imaginative policy positions voiced by Jesse Jackson. But in the campaign foggery, these ideas were no match for the orthodox fears generated by the Cold War, and the candidates lost. Still, they have one thing going for them: in the real world, these ideas make sense.

Arch Gillies is an optimist, anyway. Now he is scouting the country for a few bold men and women who might run for Congress in 1986 and who will have the nerve to articulate some of these ideas—to challenge our obsession with the Soviets head-on. "I really think it's possible," he said. "If you could find ten or fifteen people in key races who were willing to say something quite different, as different from the orthodoxy as what Reagan and conservatives were saying twenty years ago, and if some of them actually won, it would be like a lightning bolt through the system. People would say, 'My God, they got elected on that platform?'"

Ever the optimist myself, I hope he can find that kind of courage among future candidates. In the meantime, as Gillies says, "the only way to begin is to start saying these things."

Shifting to a Global Peacetime Economy

The American Committee on East-West Accord seeks to strengthen "public understanding of arms control initiatives, expanded nonmilitary trade, and mutually beneficial exchanges in science, culture, and education."* While acknowledging "fundamental differences" between the two societies, the prestigious members of the Committee, including these two authors, "see areas of mutual self-interest which, if prudently developed, can reduce tensions, advance world peace, and contribute to the well-being of both societies."

At the time the following chapter was originally published, in 1983, Cyril Black was Director of the Center for International Studies at Princeton University, and Robbin Laird was a Fellow at the Research Center on International Change at Columbia University. Believing that "it is in the American interest to support the position of Soviet moderates for trade and economic reform," they argue that a substantially higher level of Soviet trade with the West would almost force that country to change the organization of its economy in order to compete on the world market.

In contrast, Richard Pipes and some others argue that the Soviet leadership class, or *nomenklatura,* will change only under the pressure of unavoidable crisis—itself a familiar Marxist term. By giving Moscow credits, technology, and goods, Pipes argues, we would only prop up a system which, by its ideology and internal dynamics, is driven to suppress its own people and threaten the rest of the world. But while it is easy to imagine various kinds of crisis in the Soviet Union, it is less easy to specify how a crisis would lead to useful internal change rather than to suppression or war, or to say what the U.S. could do, if anything, to contribute to a salutary crisis.**

In any case, this debate deserves the closest attention. The complex argument put forth by Pipes, while not widely understood, happens to coincide in some respects with a widespread prejudice against cooperating in any way with the Soviets. In contrast, the argument in favor of trade lacks a mass constituency; there's no "gut feeling" to which it strongly appeals.

The Impact of Trade on Soviet Society

Cyril E. Black
Robbin F. Laird

Shifting to a Global Peacetime Economy

IN SEEKING TO EVALUATE THE PROBABLE IMPACT ON Soviet society in the 1980s of trade with the West and especially with the United States, it is important to assess both the changes that have taken place in international trade since the era of detente in the 1970s and the extent to which this decade of East-West trade has altered Soviet society.

In the heady days of the early 1970s when Nixon and Brezhnev ushered in the era of detente, American policy was based on the assumption that a web of interdependence in the areas of arms control, trade, and cultural, scientific and technological exchanges, would reduce the level and intensity of competition between the two countries. It was at the same time the expectation of Soviet policy that the regularization of this competitive relationship would lead to the modernization of the Soviet economy and the expansion of its global influence.

The political decision for detente led to the expansion of economic interchange. In the course of the 1970s Soviet foreign trade turnover with the West tripled, and the American share of Western exports to the Soviet Union grew from 8 percent in 1974 to 20 percent in 1979. By the early 1980s, however, while Soviet imports from the West remained high, the American share had dropped below 10 percent.

The economic downturn has been produced in part by a revised American outlook toward the Soviets in the early 1980s. The American response to the Soviet occupation of Afghanistan and to the imposition of military rule in Poland with Soviet backing led to constraints in the relations between the two countries over a whole range of dimensions, including a political decision to reduce exports to the Soviet Union. Trade is now seen as disproportionately advantageous to the Soviet Union and as contributing significantly to Soviet military capabilities.

The Soviet attitude toward the significance of foreign trade with the West has not undergone a similar change. The role formerly occupied by the United States has been replaced in part by increased trade with Japan and the countries of Western Europe.

***_Common Sense in U.S.-Soviet Trade_, edited by Margaret Chapman and Carl Marcy (American Committee on East-West Accord, 1983).**
****Richard Pipes, _Survival Is Not Enough: Soviet Realities and America's Future_ (Simon and Schuster, 1984).**

Shifting to a Global Peacetime Economy

The continuing Soviet commitment to a Western-oriented trade strategy is rooted in a fundamental change in the Soviet outlook since the days of Khrushchev. This change took the form of the evolution of Marxism-Leninism from a rather dogmatic approach which saw class conflict as the motor of history to a more pragmatic point of view which stressed instead the importance of science and technology as the basis for modern economic growth and social change. The dogmatic perspective assumed that the Soviet Union was an advanced society because the means of production had been socialized, but the pragmatists recognized that in the application of modern knowledge to development their country was not only not catching up but was steadily falling behind the United States, the countries of Western Europe, and Japan.

The new pragmatic outlook emphasizes that the U.S.S.R. needs to make effective use of science and technology. The pragmatists remain within the framework of Marxism-Leninism by maintaining that "socialist" societies are better able than "capitalist" ones to take advantage of the opportunities for human betterment offered by the revolution in science and technology. Although in due course all societies will progress from "capitalism" to "socialism," in the Soviet view, the ultimate victory of socialism is not seen in terms of class conflict or military victory. Rather the superior ability of socialism to develop and apply advanced science and technology is required. The implementation of this pragmatic perspective has led to efforts to improve production and management as well as efforts to expand foreign trade as a means both of importing advanced technology and of providing a competitive challenge to the Soviet products that must be sold to pay for the imports.[1]

Quite independent of conflicting political outlooks, economic forces have had their place as well in circumscribing the scope and nature of American-Soviet trade. In the early 1970s the confluence of a number of economic circumstances made possible the expansion of U.S.-Soviet trade. On the American side, there was a general desire to expand technology trade in order to increase the U.S. share of the global market in the face of rising Japanese and Western European competition. There was a relatively good supply of capital to be invested in foreign trade ventures. The instability of global commodity markets made access to Soviet materials attractive. On the Soviet side, there was a general desire to expand the interface of the Soviet economy with the global economy. Especially of concern to the Soviets was the modernization of the Soviet economy, to be stimulated in large part by an infusion of Western technology. A significant element of the Soviet leadership was American-oriented in the question of technology trade. Brezhnev considered the United States to be the leader in advanced technology and, hence, the logical focus of Soviet industrial trade.

By 1982 much had changed. On the American side, the capital surplus had been replaced by a capital shortfall. High interest rates had slowed significantly the flow of foreign capital in the development of large-scale investments. The economic crises of the East European countries, epitomized by Poland, increased the reluctance of Western banks to become involved in the development of the Soviet Union and the countries of Eastern Europe. On the Soviet side, there was considerable frustration with the fluctuations of American policy as well as anger over American attempts to use trade to direct political advantage. There was recognition of the relative decline of the United States in the global technology market and the increased ability of the Western European countries and Japan to provide the Soviets with the technology they seek in many areas. There was also a growing recognition of the limits of East-West trade as a stimulant to the development of the Soviet economy; such trade has not led to much-hoped-for comprehensive modernization of the Soviet economy.

The American debate on the benefits and costs of trade with the Soviet Union has been based in part on essentially economic concerns, but the effect of such trade on Soviet society has also been an important issue. The conservative perspective sees trade as strengthening the Soviet military capacity and maintains that the application of sanctions will not only halt this trend but will also induce the Soviet government to shift resources from defense to investment and consumer goods. The moderate perspective, by contrast, sees American sanctions as exerting very limited restraint on Soviet policies in view of the alternative sources of advanced technology available to it, and believes that the mutual economic benefits of trade outweigh any advantage that might be gained from sanctions.

In seeking to assess the relative validity of these two perspectives, a number of considerations based on the experience of the past decade should be taken into account. An important consideration is that the Soviet economy has become more intertwined with the global economy without becoming interdependent in the Western sense of the term. The Soviet Union is no longer economically autarkic; it trades at a level roughly comparable to other industrialized societies. The Soviet economy is also no longer insulated from global economic shocks, e.g., the importation of global inflationary pressures. But the functioning of the Soviet economy remains structurally isolated to a much greater degree than is true for other industrialized economies. Soviet managers still respond primarily to centralized plan indicators, not the demands of the global market.

A further consideration is that the Soviet Union has imported large amounts of Western technology without having effectively stimulated economic modernization. Much of the imported technology has been economically misused by the Soviets. The basic character of the system has been preserved despite the infusion of imported technology. The result has been a limited technology innovation performance, whereby the Soviets have not been able to reduce the technology gap with the West in any significant civilian area in the past decade.

A third consideration is that the importation of technology which was supposed to stimulate industrial exports was not accompanied by the entrepreneurial foresight necessary to determine the needs of the global market. The Soviets have invested in many areas where global glut, not scarcity, will prevail in the 1980s. Global industrial markets are highly driven by entrepreneurial ability; i.e., by the identification of market needs five-to-ten years in advance of the recognition by the market of those needs. The Soviet system simply lacks such an ability to operate in global markets.

It should also be noted that the Soviet Union encourages importation of information-processing equipment and the expansion of international information flows, while maintaining in large part the bureaucratic rigidity of the Soviet system. For information technology and flows to work most effectively, organizational innovation and flexibility are required. The Soviet system has been only partially rationalized from this standpoint with the creation of various types of production associations, science-production associations, science centers, etc.

The fifth consideration is the heavy emphasis placed on the significance of commercial East-West technology trade, while at the same time maintaining an important espionage effort to obtain military-related technology. Good commercial relations require trust; technological espionage undermines trust. Although, according to the National Academy of Sciences,[2] almost 70 percent of the military-related technology acquired by the Soviets in recent years has been acquired by technological espionage, American opponents of trade with the Soviets link a "hemorrhage" of military-related technology with the existence of East-West commercial activity per se. The KGB, in effect, provides indirect support for American conservatives in their efforts to suppress U.S.-Soviet trade.

An additional consideration is the Soviet emphasis on greater involvement in the global economy while maintaining as great a degree of cultural isolation as possible. Much of the intellectual ferment of the 1970s can be laid at the feet of the Soviet opening to the West. The Soviets have responded both by suppression of political dissent and by expanding the effort to make Soviet science and technology an integral part of global scientific and technological development. There is a significant contradiction between economic, scientific, and technological interdependence and culturally enforced isolation.

One should also consider the fact that the scope and nature of foreign trade have increased without any relaxation in the primacy of the Ministry of Foreign Trade. The expansion of Soviet trade activity has led to some modification of the Soviet foreign trade system; there has been greater involvement by organizations other than the Ministry of Foreign Trade (State Committee on Science and Technology, industrial associations, etc.). Nonetheless, the participation of a greater number of organizations has not led to a reduction of the Ministry's control over purchasing decisions. Maintaining the monopoly of the Ministry of Foreign Trade undercuts the effort to improve the efficiency of the Soviet economy.

It is also significant that restrictive political control over the countries of Eastern Europe has been increased, while they have been encouraged at the same time to expand their economic interchange with the West. The expansion of East European trade with the West has stimulated economic rationalization and reform efforts in Eastern Europe, and such efforts have an indirect effect upon the Soviet economic system as well.

In light of the ambiguous impact which trade has had on the Soviet system, including the existence of differences of opinion within the Soviet elite itself, what might be considered to be the American interests in Soviet-American trade in the 1980s?

First, it is not in the American interest to support the position of Soviet conservatives in their attempt to close off the Soviet Union and its allies from outside influence, including trade relations. It is in the American interest to support the position of Soviet moderates for trade and economic reform.

Second, it is not in the American interest to provide the Soviet Union with goods directly useful for Soviet military purposes. The U.S. government has a legitimate interest in ensuring that such trade does not take place. But it is in the American interest to have minimal government interference in nonstrategic trade areas so that the business organizations which are the most capable of determining the profitability of trade are able to do so.

The Soviet armament program reflects the perception of Soviet leaders that they face national security threats from many quarters. Most expert observers believe the Soviet arms program is going well beyond reasonable limits and represents a potent threat that calls for serious countermeasures. The way to meet this challenge is not by reducing trade, however, but by taking sound military measures including comprehensive strategic arms reduction negotiations designed to bring the problem under control.

Third, it is not in the American interest to turn the East European and Soviet markets over to its industrial competitors. It is in the American interest to have a relatively depoliticized trade policy to encourage as stable trade relationships as possible. Without more predictability in our trade relationships, American trade with the East will not grow.

Fourth, it is not in the American interest to favor the exclusion of the Soviet Union and its allies from international organizations such as the International Monetary Fund. The United States should encourage the normalization of economic relations within an IMF framework.

Fifth, it is not in the American interest to encourage the use of trade for clearly unrelated political goals. Expanded trade should be linked with the normalization of economic interchange, with increased financial disclosure, with the development of price reforms, and with the modifications of the foreign trade monopoly—all of which are political changes legitimately related to normalization of trade matters.

In short, American-Soviet trade as a strand in the expanded web of involvement of the Soviet economy in the global system can nurture domestic change in the Soviet Union. But the changes to be anticipated will be only steps in the gradual opening up of the Soviet Union to global society and of less confrontationist policies toward the West. Trade relations cannot be expected by themselves to form the basis for an enduring peace between the superpowers; political and cultural efforts are needed as well. Trade relations cannot ensure peace, but they can contribute to the process of building peace.

1. The scope and significance of this ideological transformation are discussed in Erik P. Hoffman and Robbin F. Laird, *The Politics of Economic Modernization in the Soviet Union* (Ithaca, NY: Cornell University Press, 1982), and Erik P. Hoffman and Robbin F. Laird, *The Scientific-Technological Revolution and Soviet Foreign Policy* (New York: Pergamon Press, 1982).
2. Report on "Scientific Communication and National Security," National Academy of Sciences, National Academy Press, Washington, D.C., Fall 1982.

What's it like to engage in a joint venture in the Soviet bloc? How do you deal with your partners? Ask William Norris, founder of Control Data. "One of the best ways I know to evaluate people," says this Minnesota resident, "is to go fishing with them." In the course of the relationships that he describes in this chapter, there was many a cast without a nibble, the line sometimes got caught in weeds, and a big one got away as it was being pulled toward the gunwales. But in the author's view, no harm was done and the partners found some fish to fry. When the politicians are ready to issue more licenses, Norris and many of his competitors will be ready to go out again.

Joint Ventures with the Eastern Bloc

William C. Norris

FOR THE PAST 20 YEARS, I HAVE BEEN A STRONG ADVOCATE of expanding trade and scientific cooperation with the Soviet Union and Eastern Europe, and Control Data has made a substantial investment in developing such relationships with a number of those countries. In the process, we have accumulated a vast amount of experience and proven what we postulated initially, namely that there was great potential for profitable business relationships between the East and West based on long-term cooperative ventures.

More specifically, our experience has shown that normally, these relationships should not be based on a policy of just licensing technology. Rather, any technology transfer should firmly rest on the principle of receiving equivalent value in return. The best way, then, is to receive *technology* in return which is deemed to have equivalent value, but it could be achieved also by obtaining a share of market or by some combination of the two. The preferred method of implementing this policy is a cooperative venture where the partners have approximately equal participation. . . .

Experience is always the best teacher. Let me describe some of the lessons we have learned as a result of our experience with Soviet Union and Romanian programs. First, I will talk about a successful Control Data-Romania joint venture, ROM-Control Data, a company which is 45 percent owned by Control Data and 55 percent owned by the Government of Romania. ROM-Control Data opened its doors in 1973 after a two-year preparatory effort.

During that time, persons in the United States with a positive attitude toward such an activity were scarce as hen's teeth, whereas the woods were full of doubters. A common admonition was that "those communists will confiscate the plant."

Because of the dire predictions, I decided to visit Romania to meet with a number of Romanian government officials. One of the best ways I know to evaluate people is to go fishing with them. I've never known an untrustworthy person who liked to fish, so just the acceptance of an invitation to fish is a positive sign. Sure enough, one of the ministers accepted my invitation, and we fished the Danube Delta.

It was an enjoyable, memorable and rewarding experience. Scenic places comparable to the Danube Delta are rare. It was my first and only time to use a hydrofoil cruiser for fishing. Because of its high speed and low draft, we were able to reach remote areas where the fishing was best in a short time. During the trip, I became much better acquainted with our potential joint venture partners and gained confidence that they would fulfill their obligations in a joint venture.

That turned out to be an accurate judgment. Ironically, the most serious problems were generated on our side by export controls. The many unusual problems arising from dissimilar languages and political and economic systems, however, were resolved. On the other hand, export control problems developed into a plague on the enterprise which at times threatened its viability.

The initial products manufactured by ROM-Control Data were punched card and printer peripheral equipment for computers. They were older models for which demand in the U.S. and Europe was declining because of availability of higher-performance, more cost-effective equipment. It has never been possible to get export licenses for products embodying state-of-the-art technology—even obtaining licenses for older equipment has required protracted negotiations.

Most recently, a license application for the transfer of two magnetic disc products embodying a ten-year-old technology was in process for a year. In spite of no evidence that any of the parts or equipment in the plant had ever been diverted from authorized use, the U.S. government imposed stricter accounting of inventory, to which we agreed. Then the requirement was added to maintain Control Data employees resident in Bucharest to keep track of how many parts came into the plant and how many finished units went out, 24 hours a day, 7 days a week. Compliance with the employee residency requirement would have cost $400,000. This added cost made the transaction financially unsound. Meanwhile, with the year's delay in processing the application, the window of market opportunity had narrowed to the point where the two products became economically unfeasible. Hence, the applications were withdrawn and export licenses for alternative products are being applied for.

This and other license application experiences demonstrate the inherent conservative bias in the U.S. export control system engendered by bureaucratic pressures. Equally bad are occasional erratic decisions apparently caused by idealogical considerations.

Despite the many export-control and other problems, ROM-CDC has developed into a solid company. Annual sales have reached a level of $12 million; the company is profitable and pays annual dividends. In fact, the initial cash investment by Control Data has been repaid. Most important, ROM-Control Data has commenced its own product research and development.

Also gratifying is the excellent rapport which has developed between Control Data and the Romanian government. Significant confirming evidence of it surfaced during the visit of President Ceaucescu to the U.S. when he offered to allow Control Data to increase its share of ownership from 45 percent to 55 percent. I declined because I'm a great believer that when anything is working well don't tinker with it.

My only regret with respect to ROM-CDC is that without the unduly restrictive export control policies and interminable delays in their administration, the company would be five times its present size with a corresponding increase in jobs, both in Romania and the U.S.

Turning now to the Soviet Union, it was the umbrella agreements signed at the 1972 and 1973 summit meetings between the U.S. and that country which provided the initial encouragement for U.S. companies to pursue cooperative ventures there. Having laid the groundwork in the Soviet Union during the preceding four years, we at Control Data Corporation were able to move rapidly to conclude a broad-based ten-year frame agreement in November, 1973, with the state committee of the USSR Council of Ministries for Scientific and Technological Cooperation.

The Control Data agreement was responsive to the objectives of the umbrella agreements of 1972 and 1973 to facilitate technological cooperation and expansion of trade. Soviet objectives included those of obtaining technology to manufacture process control systems and computer peripheral equipment and the purchase of computers for scientific computing, seismic exploration, weather forecasting, and education. Control Data's expectations included a share of the market and technology in applying computers in process control, health care, education, and scientific computing, and an opportunity to increase greatly the number of U.S. companies engaging in commercializing results of Soviet research. The Control Data agreement was the broadest in scope of any frame agreement between U.S. companies and the Soviet Union, of which there were 58 by 1975.

Mutual objectives were consolidated into specific projects which included the development of a new type of computer, development and operation of a computer communications network, sale of computer systems, development and manufacture of peripheral equipment, transfer and commercialization of selected Soviet technologies, and development of computer application software in the areas of health care, education, process control, and scientific problem solving.

Overall management of the frame agreement was accomplished by a coordinating committee with five people from each side. Working groups were established to define and plan projects in each area of cooperation.

Until the June, 1975 meeting of the coordinating committee, progress seemed to be agonizingly slow; however, a breakthrough occurred at that meeting. Final specifications on major projects were agreed upon. Most important, the projects were included in the 1976–1980 Soviet five-year plan, thus meeting the schedule initially set up for achieving this necessary step.

By the time work could commence on individual projects, the enthusiasm in the U.S. following the general 1972 U.S.-Soviet agreement on scientific and technological cooperation was starting to wane. As a result, we were faced with increasing difficulty in getting export licenses, and the regulations were often capriciously administered.

No better example of the barriers to consistent implementation of East-West trade policy is the story of the hydrometeorological research center in Moscow. Early on, both the U.S. and the Soviet Union had identified weather as a potential arena for technological cooperation. Both countries were and are members of the World Meterological Organization (W.M.O.). Information and weather data are shared among all 144 members of the W.M.O. Around the world there are some 150 weather centers and 22 of these have the burden of collecting and processing the data collected in their area and sharing this around the world. Only the center in Moscow lacked the necessary large-scale computer capability; and because of this, there was a serious gap in the information because the vast amount of data gathered in Siberia could not be processed in enough time.

The computer in use by W.M.O. throughout the 1970s was the Control Data 7600. Thus, there came to be a proposal for such a system at Hydromet in Moscow.

At about this period, columnist Jack Anderson learned about the license application and wrote an inflammatory column denouncing the sale. Opening sentences read, "Control Data is preparing to sell the Soviets a $13 million electronic brain which could be turned against us to track U.S. missiles, planes and submarines. It is also capable of decoding sensitive U.S. intelligence transactions. The miracle machine is the Cyber 7600 which will soon be on its way to the Soviet Union unless there is a last minute stop order." Of course, a single 7600 computer cannot be turned against us in a meaningful way to track U.S. missiles, planes and submarines, but the average person doesn't know that. Nor, of course, was there any mention of the real reason for the sale.

Very quickly, the White House and several members of Congress were competing to get the credit for stopping the sale. In spite of efforts by a number of additional Congressmen to get a thorough and objective review, that opportunity was never available, and the license was denied—primarily on an emotional basis instead of a rational careful analysis.

Our Soviet partners were stunned and their commitment to our cooperative effort began to decline. With the continuing deterioration in overall U.S.-Soviet relationships, progress slowed and all work eventually ceased.

The only project which progressed far enough to yield meaningful results was in the area of technology transfer. This project commenced with a survey of commercializable technology by our Worldtech division, which is a worldwide organization for assisting in the transfer of technologies through licensing arrangements or cooperative ventures. Associated with Worldtech is Technotec, which is a data base of available or wanted technologies.

Because of the broad scope of the frame agreement, we had access to a large number of Soviet laboratories and plants. In fact, we were told that Control Data personnel visited more Soviet research, engineering, and manufacturing facilities than employees with any other corporation.

A list of 340 technologies suitable for transfer was assembled. Since technology can't be successfully transferred overnight, there was only enough time to transfer two technologies before the work ended on all projects. One was a technology embodying the most efficient design of a machine for crushing stone. A license for this technology was sold to a large U.S. company.

The second technology was a process using titanium nitride for hardening machine tools. This technology was sold to a new company, called Multiarc, which Control Data helped establish and in which we have almost 40 percent ownership.

Multiarc is a very successful company with a market value today of nearly $60 million. Since the cost of our 40 percent interest was less than $3 million, we have an unrealized profit of some $20 million, which is far in excess of the cost incurred by us in the other cooperative projects which had to be abandoned without having achieved results.

Multiarc's success illustrates the advantages of a cooperative venture over the sale of technology. While it wasn't feasible at the time of the transaction to consider the alternative of a cooperative venture, obviously, the Soviet Union would have gained more financially from being a partner in such an approach. In addition, cooperation would have been likely between the Soviet researchers in that field and Multiarc, which would have created technology for product improvements and for new cooperative projects.

Our successful consummation of two technology transfers—one highly profitable—and our broad view of technology in the Soviet Union further confirms the enormous potential for U.S. companies to commercialize Soviet research results through cooperative ventures.

That this is true is to be expected considering the vast research effort in the Soviet Union, where there are more scientists engaged in research than in the U.S., and the fact that the Soviet Union has a large number of the best scientists in many fields—theoretical physics and mathematics, to name two.

What the Soviet Union lacks is a marketing capability to efficiently commercialize its research. Marketing is a major strength of U.S. companies; hence, the great opportunity for mutually rewarding relationships through cooperative ventures.

Having described some of our experience with cooperative programs with Romania and the Soviet Union and outlined the potential for cooperation, let me now review the advantages of cooperative ventures for developing business with socialist countries.

Participants should have approximately equal ownership. Preferably, the cooperative venture should have both manufacturing and marketing capabilities. In addition, R&D should be carried on to make improvements in the products and eventually develop new ones.

High on the list of advantages of a cooperative venture is that the technology involved is fairly valued because that is established by the marketplace. Where values are established by negotiation for the sale of a license, one side or the other is invariably short-changed.

Through a cooperative venture, it is possible to share permanently in socialist country markets, whereas, with license, the relationship is usually fairly short. In addition, a cooperative venture provides an adequate degree of control over the technology.

It is also practicable, even desirable, in a cooperative venture located abroad to retain some of the component production and related R&D in the United States. The cooperative venture as well as both owners benefit from production and R&D carried on in both countries.

Concerns about the possible incompatibility between cooperative ventures and centrally planned socialist economies can be addressed because such ventures are inherently flexible enough to accommodate to the requirements of both capitalist and socialist economies. For example, if the tenets of the socialist system in a particular country requires 51 percent ownership, this can be accommodated with management control retained, if indicated, by contractual methods. Restrictions on the ownership of land, buildings, and equipment can be met through leasing.

Probably the most important advantage of a cooperative venture is the glue it provides to keep the participants working together during periods of adversity. Having invested significant time and resources and made widely visible commitments to achieve successful operation, both participants regard the resolution of problems, even very difficult ones, as much preferable to failure.

Finally, the interrelationships of employees in a cooperative venture in a socialist country with those of the U.S. partner builds mutual respect and understanding which is so badly needed. The main point is that both nations, however different, need to adapt their approaches to facilitate economic, scientific, and technological cooperation.

If cooperative ventures between the U.S. and Soviet bloc countries are to proliferate and reach their great potential for increasing East-West trade and providing bridges for better relationships, we will have to forego the use of trade sanctions to achieve foreign policy and national security objectives.

Cooperative ventures require substantial and long term commitments. Changes in government policies effecting the transfer of financial or technical resources can severely disrupt such operations; hence, U.S. coporations are at present reluctant to enter into cooperative ventures with socialist countries. There is also a growing concern among European companies engaging in cooperative ventures with U.S. companies even when they are located outside the United States because of the extraterritorial reach of U.S. export controls as demonstrated by the Soviet pipeline sanctions. . . .

Evidence is mounting that U.S. policies haven't achieved their objectives. The Soviet military doesn't appear to have been materially disadvantaged, and business opportunities denied U.S. companies have invariably been snapped up by other countries.

Legislation establishing appropriate policies, although urgently needed, isn't going to be put in place in the near future because of a number of obstacles. One of the major obstacles is the low level of understanding of technology and its implications by a high percentage of our population. Too often, everything is placed under the category of technology from research to products. Of course, this is incorrect. Technology is know-how—the application of the results of research to producing products and services. In addition, technology is frequently characterized as being high, advanced, or sophisticated without adequately defining what is meant. This causes the average person to be uneasy because of being unable to evaluate the significance of equipment sales, technology transfers, and cooperative research programs.

Even though many people see the Soviet Union as a competitor and adversary, few—including President Reagan—are in favor of economic warfare. On the other hand, there is a reluctance—usually unspoken—on the part of some to do business at all because they don't want to see the Soviet economy strengthened. Whenever a business transaction is proposed, reasons are sought for opposing it—often on the basis that military capability will be enhanced—because virtually any technology has military application to one extent or another. This has led to extensive controls, which have kept trade with the Soviet Union at a low level, closed the doors on access to important technologies in that country, and caused concerns by West Europe about being able to gain the full benefits from cooperative research with U.S. organizations.

What is needed is a reduction in controls as recently recognized by the Presidential Commission on Industrial Competitiveness, which said, "It is important that we make it clear that national security and foreign policy are dependent on maintaining our industrial competitiveness. Export restrictions are generally antithetical to U.S. competitiveness because they posture U.S. industries as unreliable suppliers." In addition, we should move in the direction that will assure that in any transaction with the Soviet Union transferring important technology, we will get back an equivalent technology, and we should widely publicize that information.

Even more important is that the average person doesn't understand that most new jobs in our country are derived from the process of applying technology. As I noted previously in discussing the hundreds of technologies suitable for transfer that we found in our visits to Soviet laboratories and plants, cooperative ventures have an enormous potential for creating jobs. In addition to the two examples I cited earlier, further evidence of the potential is provided by the more than 50 licenses for the transfer of Soviet technologies to American companies which were signed since 1971. These technologies were from a number of fields including medical and surgical procedures and apparatus, metal casting, welding, chemical engineering, and machine tools. . . .

5 Cooperating in Outer Space

Orbiting the earth, the astronauts taught us to see it as a whole. On a schoolroom globe the various "nations" had distinct boundaries, but from space there were only a complex swirl of clouds, the enormous oceans, and khaki continents colored green, in patches, by chlorophyll. Where people gather in cities there was a twinkle at night, partially obscured by smog. The earth was revealed as a spherical organism, a very thin culture of life growing on a globe.

When the first color photographs were published showing the earth from space, I used to dream about seeing it as the astronauts did. If the film of life were a pond, I wanted to experience it the way an eagle does, not only as a fish. When I was invited to have lunch with Rusty Schweikart, one of the Apollo astronauts, I mentioned this image to him. "When you're in Washington," he replied, "go see *The Dream Is Alive* at the NASA museum." Shot in IMAX, a new 70 mm film process, it was projected on an enormous screen. "It's ninety-five percent as good as being in space," said Rusty. "What's the other five?" I asked. "Zero gravity," he replied.

I saw the film before the Challenger explosion, when the dream was more alive than it now seems. The shot that made the deepest impression on me was an easy pan starting in Northern Italy, moving down to the "boot," and continuing over the Nile delta. We had sliced across much of an ancient empire in the time it would take to eat a snack. It's a shame that Shakespeare didn't have this footage for the *Cleopatra* credits. In any case, I was so taken with the film—allowing myself to become what Gordon Feller calls an "intronaut"—that I didn't begrudge Rusty his experience of zero gravity. As people in ancient times circumambulated a city's walls or a sacred peak, as they later crossed a thin strip of territory by rail, so our space-explorers are now holding whole continents in their awareness.

So far there are no mega-weapons in space. Nuclear submarines prowl the sea; missiles are installed in "silos" under the wheat fields and tundra; and bombers wait at the end of runways. But in space there are not yet any hypervelocity projectile accelerators, ion beam accelerators, chemical or eximer lasers, or any of the other devices being studied by America's Strategic Defense Initiative Organization. These devices are concepts, lab experiments, or mock-ups. They are not above our heads yet. The globe is not yet enshrouded in what would be an enormous, spherical, fast-moving and lethal video-game.

President Reagan says he wants to render all nuclear weapons "impotent and obsolete." As soon as the U.S. develops a working "Star Wars" system, a successor to Reagan would supposedly sell a replica to the Soviets "at cost," so they could blow up our rockets as we'd be able to blow up theirs. Considering the enormous *estimated* cost of this system, the Pentagon's record of extensive cost over-runs, and our ability to foil any system that we design, this is a droll offer. Instead of selling them the system, why not just auction off everybody's missiles to be destroyed? That way there'd be nothing to defend against.

The ridiculous and tragic nature of the "Star Wars" project is revealed in the following chapters by Bundy, Kennan, McNamara, and Smith (who are collectively known, in some circles, as the four wise men) and by Mische, a wise woman. One of the few good arguments for the project is that it could be traded away for a Soviet agreement to make deep cuts in the weapons against which it would seek to defend. However, while many a "bargaining chip" has been developed, few have actually been negotiated away. In arms control negotiations, an observer has said, a bargaining chip is a weapon built to mollify members of the Senate and the Joint Chiefs of Staff, in return for their supporting a treaty that forbids something else that they don't necessarily want to build anyway.

The President gained enormous support for his Strategic Defense Initiative by arguing that the present system, "mutual assured destruction," is immoral, and that it's better to defend our cities, industries, and people against an attack, preferably by using non-nuclear means, than to threaten the other side with sudden ghastly destruction. He's right.[1] The only trouble is, there's no prospect of actually being able to defend anything except some missile silos—in order to make sure we can retaliate, which returns us to the theme of "assured destruction." And if we were to make a technical breakthrough on missile defense, it would so terrify the Soviets they might do *anything.*

However, the *ambition* to create a defense, the dream of doing it, is so attractive that the President's proposal immediately shifted the moral terms of the debate. Before he made it, the proponents of big military budgets could be charged with "accelerating the arms race"; now they could reply they wanted only to find a defense, to make all those big missiles "impotent." It was the arms control crowd that had left the country totally exposed to retaliation, on account of its treasured Anti-Ballistic Missile Treaty, its ban on space weapons, its urgency to stop the nuclear testing needed to develop ray-guns to blow up nuclear missiles. The prospect of creating a defense allowed the President to ask the defenders of "mutual assured destruction" whether they expected the system of deterrence would work without a hitch forever.

The public does not believe it will. In one especially sophisticated survey, thirty-eight percent of Americans said the chances of a nuclear war in the next decade are (very or fairly) great.[2] Since the survey was conducted in 1984, the respondents were talking about 1994. A child enrolled in kindergarten as I write will not have graduated from junior high school by then. If offered a choice between a precarious system of mutual deterrence, without any defense, and a "shield" against an attack, however it occurs, who would not opt for the shield?

I leave it to the next two chapters to explain what's wrong with the plan for a so-called shield. Here it needs to be noted that space is already being "used," not for positioning weapons but for watching the planet. Of the "national technical means" used to verify compliance with arms control, the most important means are in orbit. True, satellites also serve as a "force-multiplier," in that, as long as they are working, they can track moving targets, guide missiles more exactly, assess damage, and carry an enormous flow of military command data. The same technology that keeps count of strategic forces, allowing each side to know that the other is respecting the limits, also amplifies the destructive potential of weapons on earth. The question is how we will use the technology.

We could use it, for example, on behalf of an International Satellite Monitoring Agency.

Imagine we are in orbit, looking down at our one globe. In the superpower capitals far below, leaders are explaining why we have to spend our wealth on creating a defense that both sides will acquire, in order to prevent either from destroying the other. Wouldn't it be much safer, quicker, and cheaper to get rid of the existing weapons we're all afraid of? We could then use space not as a battle zone but as a mirror for compliance.

—Craig Comstock

1. For two of the most recent moral analyses, see Anthony Kenny, *The Logic of Deterrence* (University of Chicago Press, 1986) and Joseph S. Nye, Jr., *Nuclear Ethics* (The Free Press, 1986).
2. The Public Agenda Foundation in collaboration with the Center for Foreign Policy Development at Brown University, *Voter Options on Nuclear Arms Policy* (1984), page 22.

Are these authors qualified? McGeorge Bundy was special assistant to Presidents Kennedy and Johnson for national security affairs, president of the Ford Foundation from 1966 to 1979, and is now professor of history at New York University. George F. Kennan, now professor emeritus at the Institute for Advanced Study, was ambassador to the Soviet Union in 1952 and to Yugoslavia in 1961–63 and author of many works of diplomatic history. He was also the originator of an approach to "containment" that was more flexible and subtle than the policy adopted under that name. Robert S. McNamara was secretary of defense from 1961 to 1968, then president of the World Bank until 1981. Gerard Smith was chief of the U.S. delegation to the Strategic Arms Limitation Talks (SALT) from 1969 to 1972, and is author of *Doubletalk: The Story of SALT I.*

Joining forces, these four offer advice to the President, as they have each been asked to do many times in the past. In the case of President Reagan, they request him to reconsider his commitment to the Strategic Defense Initiative and to seek arms control instead. Their paper originally appeared in *Foreign Affairs.*

The President's Choice

McGeorge Bundy
George F. Kennan
Robert S. McNamara
Gerard Smith

I

THE REELECTION OF RONALD REAGAN MAKES THE FUTURE of his Strategic Defense Initiative the most important question of nuclear arms competition and arms control on the national agenda since 1972. The President is strongly committed to this program, and senior officials, including Secretary of Defense Caspar W. Weinberger, have made it clear that he plans to intensify this effort in his second term. Sharing the gravest reservations about this undertaking, and believing that unless it is radically constrained during the next four years it will bring vast new costs and dangers to our country and to mankind, we think it urgent to offer an assessment of the nature and hazards of this initiative, to call for the closest vigilance by Congress and the public, and even to invite the victorious President to reconsider. While we write only after obtaining the best technical advice we could find, our central concerns are political. We believe the President's initiative to be a classic case of good intentions that will have bad results because they do not respect reality.

This new initiative was launched by the President on March 23, 1983, in a surprising and quite personal passage at the end of a speech in praise of his other military programs. In that passage he called on our scientists to find means of rendering nuclear weapons "impotent and obsolete." In the briefings that surrounded the speech, Administration spokesmen made it clear that the primary objective was the development of ways and means of destroying hostile missiles—meaning in the main Soviet missiles—by a series of attacks all along their flight path, from their boost phase after launch to their entry into the atmosphere above the United States. Because of the central position the Administration itself gave to this objective, the program promptly acquired the name Star Wars, and the President's Science Advisor, George Keyworth, has admitted that this name is now indelible. We find it more accurately descriptive than the official "Strategic Defense Initiative."[1]

II

What is centrally and fundamentally wrong with the President's objective is that it cannot be achieved. The overwhelming consensus of the nation's technical community is that in fact there is no prospect whatever that science and technology can, at any time in the next several decades, make nuclear weapons "impotent and obsolete." The program developed over the last 18 months, ambitious as it is, offers no prospect for a leak-proof defense against strategic ballistic missiles alone, and it entirely excludes from its range any effort to limit the effectiveness of other systems—bomber aircraft, cruise missiles, and smuggled warheads.

The President's hopes are entirely understandable. There must be very few Americans who have never shared them. All four of us, like Mr. Reagan, grew up in a world without nuclear weapons, and we believe with passion that the world would be a much safer place without them. Americans should be constantly on the alert for any possibilities that can help to reduce the nuclear peril in which we all live, and it is entirely natural that a hope of safety like the one the President held out should stir a warmly affirmative first response. But false hope, however strong and understandable, is a bad guide to action.

The notion that nuclear weapons, or even ballistic missiles alone, can be rendered impotent by science and technology is an illusion. It reflects not only technological hubris in the face of the very nature of nuclear weapons, but also a complete misreading of the relation between threat and response in the nuclear decisions of the superpowers.

The first and greatest obstacle is quite simply that these weapons are destructive to a degree that makes them entirely different from any other weapon in history. The President frequently observes that over the centuries every new weapon has produced some countervailing weapon, and up to Hiroshima he is right. But conventional weapons can be neutralized by a relatively low rate of kill, provided that the rate is sustained over time. The classic modern example is defense against non-nuclear bombing. If you lose one bomber in every ten sorties, your force will soon be destroyed. A pilot assigned to fly 30 missions will face a 95-percent prospect of being shot down. A ten-percent rate of kill is highly effective.

With nuclear weapons the calculation is totally different. Both Mr. Reagan's dream and his historical argument completely neglect the decisive fact that a very few nuclear weapons, exploding on or near population centers, would be hideously too many. At today's levels of superpower deployment—about 10,000 strategic warheads on each side—even a 95-percent kill rate would be insufficient to save either society from disintegration in the event of general nuclear war. Not one of Mr. Reagan's technical advisers claims that any such level of protection is attainable. They know better. In the words of the officer in charge of the program, Lieutenant General James Abrahamson, "a perfect defense is not a realistic thing." In response to searching questions from Senator Sam Nunn of Georgia, the senior technical official of the Defense Department, Under Secretary Richard DeLauer, made it plain that he could not foresee any level of defense that would make our own offensive systems unnecessary.

Among all the dozens of spokesmen for the Administration, there is not one with any significant technical qualifications who has been willing to question Dr. DeLauer's explicit statement that "There's no way an enemy can't overwhelm your defenses if he wants to badly enough." The only senior official who continues to share the President's dream and assert his belief that it can come true is Caspar Weinberger, whose zealous professions of confidence are not accompanied by technical support.

The terrible power of nuclear weapons has a second meaning that decisively undermines the possibility of an effective Star Wars defense of populations. Not only is their destructive power so great that only a kill rate closely approaching 100 percent can give protection, but precisely because the weapons are so terrible neither of the two superpowers can tolerate the notion of "impotence" in the face of the arsenal of the opponent. Thus any prospect of a significantly improved American defense is absolutely certain to stimulate the most energetic Soviet efforts to ensure the continued ability of Soviet warheads to get through. Ever since Hiroshima it has been a cardinal principle of Soviet policy that the Soviet Union must have a match for any American nuclear capability. It is fanciful in the extreme to suppose that the prospect of any new American deployment which could undermine the effectiveness of Soviet missile forces will not be met by a most determined and sustained response.

This inevitable Soviet reaction is studiously neglected by Secretary Weinberger when he argues in defense of Star Wars that today's skeptics are as wrong as those who said we could never get to the moon. The effort to get to the moon was not complicated by the presence of an adversary. A platoon of hostile moon-men with axes could have made it a disaster. No one should understand the irrelevance of his analogy better than Mr. Weinberger himself. As secretary of defense he is bound to be familiar with the intensity of our own American efforts to ensure that our own nuclear weapons, whether on missiles or aircraft, will always be able to get through to Soviet targets in adequate numbers.

The technical analyses so far available are necessarily incomplete, primarily because of the very large distance between the President's proposal and any clearly defined system of defense. There is some truth in Mr. Weinberger's repeated assertion that one cannot fully refute a proposal that as yet has no real content. But already important and enduring obstacles have been identified. Two are systemic and ineradicable. First, a Star Wars defense must work perfectly the very first time, since it can never be tested in advance as a full system. Second, it must be triggered almost instantly, because the crucial boost phase of Soviet missiles lasts less than five minutes from the moment of launch. In that five minutes (which new launch technology can probably reduce to about 60 seconds), there must be detection, decision, aim, attack and kill. It is hard to imagine a scheme further removed from the kind of tested reliability and clear presidential control that we have hitherto required of systems involving nuclear danger.

There are other more general difficulties with the President's dream. Any remotely leak-proof defense against strategic missiles will require extensive deployments of many parts of the system in space, both for detection of any Soviet launch and, in most schemes, for transmission of the attack on the missile in its boost phase. Yet no one has been able to offer any hope that it will ever be easier and cheaper to deploy and defend large systems in space than for someone else to destroy them. The balance of technical judgment is that the advantage in any unconstrained contest in space will be with the side that aims to attack the other side's satellites. In and of itself this advantage constitutes a compelling argument against space-based defense.

Finally, as we have already noted, the President's program offers no promise of effective defense against anything but ballistic missiles. Even if we assume, against all the evidence, that a leak-proof defense could be achieved against these particular weapons, there would remain the difficulty of defense against cruise missiles, against bomber aircraft, and against the clandestine introduction of warheads. It is important to remember here that very small risks of these catastrophic events will be enough to force upon us the continuing need for our own deterrent weapons. We think it is interesting that among the strong supporters of the Star Wars scheme are some of the same people who were concerned about the danger of the strategic threat of the Soviet Backfire bomber only a few years ago. Is it likely that in the light of these other threats they will find even the best possible defense against missiles a reason for declaring our own nuclear weapons obsolete?

Inadvertent but persuasive proof of this failing has been given by the President's science adviser. In February 1984, in a speech in Washington, Mr. Keyworth recognized that the Soviet response to a truly successful Star Wars program would be to "shift their strategic resources to other weapons systems," and he made no effort to suggest that such a shift could be prevented or countered, saying: "*Let* the Soviets move to alternate weapons systems, to submarines, cruise missiles, advanced technology aircraft. Even the critics of the President's defense initiative agree that *those* weapons systems are far more stable deterrents than are ICBMs [land-based missiles]." Mr. Keyworth, in short, is willing to accept all these other means of warhead delivery, and he appears to be entirely unaware that by this acceptance he is conceding that even if Star Wars should succeed far beyond what any present technical consensus can allow us to believe, it would fail by the President's own standard.

The inescapable reality is that there is literally no hope that Star Wars can make nuclear weapons obsolete. Perhaps the first and most important political task for those who wish to save the country from the expensive and dangerous pursuit of a mirage is to make this basic proposition clear. As long as the American people believe that Star Wars offers real hope of reaching the President's asserted goal, it will have a level of political support unrelated to reality. The American people, properly and sensibly, would like nothing better than to make nuclear weapons "impotent and obsolete," but the last thing they want or need is to pay an astronomic bill for a vastly intensified nuclear competition sold to them under a false label. Yet that is what Star Wars will bring us, as a closer look will show.

III

The second line of defense for the Star Wars program, and the one which represents the real hopes and convictions of both military men and civilians at the levels below the optimistic President and his enthusiastic secretary of defense, is not that it will ever be able to defend *all our people*, but rather that it will allow us to defend *some of our weapons and other military assets*, and so, somehow, restrain the arms race.

This objective is very different from the one the President has held out to the country, but it is equally unattainable. The Star Wars program is bound to exacerbate the competition between the superpowers in three major ways. It will destroy the Anti-Ballistic Missile (ABM) Treaty, our most important arms control agreement; it will directly stimulate both offensive and defensive systems on the Soviet side; and as long as it continues it will darken the prospect for significant improvement in the currently frigid relations between Moscow and Washington. It will thus sharpen the very anxieties the President wants to reduce.

As presented to Congress in March 1984, the Star Wars program calls for a five-year effort of research and development at a total cost of $26 billion. The Administration insists that no decision has been made to develop or deploy any component of the potential system, but a number of hardware demonstrations are planned, and it is hoped that there can be an affirmative decision on full-scale system development in the early 1990s. By its very nature, then, the program is both enormous and very slow. This first $26 billion, only for research and development, is not much less than the full procurement cost of the new B-1 bomber force, and the timetable is such that Mr. Reagan's second term will end long before any deployment decision is made. Both the size and the slowness of the undertaking reinforce the certainty that it will stimulate the strongest possible Soviet response. Its size makes it look highly threatening, while its slowness gives plenty of time for countermeasures.

Meanwhile, extensive American production of offensive nuclear weapons will continue. The Administration has been at pains to insist that the Star Wars program in no way reduces the need for six new offensive systems. There are now two new land-based missiles, two new strategic bombers, and two different submarine systems under various stages of development. The Soviets regularly list several other planned American deployments as strategic because the weapons can reach the Soviet homeland. Mr. Reagan recognized at the very outset that "if paired with offensive systems," any defensive systems "can be viewed as fostering an aggressive policy, and no one wants that." But that is exactly how his new program, with its proclaimed emphasis on both offense and defense, is understood in Moscow.

We have been left in no doubt as to the Soviet opinion of Star Wars. Only four days after the President's speech, Yuri Andropov gave the Soviet reply:

On the face of it, laymen may find it even attractive as the President speaks about what seem to be defensive measures. But this may seem to be so only on the face of it and only to those who are not conversant with these matters. In fact the strategic offensive forces of the United States will continue to be developed and upgraded at full tilt and along quite a definite line at that, namely that of acquiring a first nuclear strike capability. Under these conditions the intention to secure itself the possibility of destroying with the help of the ABM defenses the corresponding strategic systems of the other side, that is of rendering it unable of dealing a retaliatory strike is a bid to disarm the Soviet Union in the face of the U.S. nuclear threat.[2]

The only remarkable elements in this response are its clarity and rapidity. Andropov's assessment is precisely what we should expect. Our government, of course, does not intend a first strike, but we are building systems which do have what is called in our jargon a prompt hard-target kill capability, and the primary purposes of these systems is to put Soviet missiles at risk of quick destruction. Soviet leaders are bound to see such weapons as a first-strike threat. This is precisely the view that our own planners take of Soviet missiles with a similar capability. When the President launches a defensive program openly aimed at making Soviet missiles "impotent," while at the same time our own hard-target killers multiply, we cannot be surprised that a man like Andropov saw a threat "to disarm the Soviet Union."[3] Given Andropov's assessment, the Soviet response to Star Wars is certain to be an intensification of both its offensive and defensive strategic efforts.

Perhaps the easiest way to understand this political reality is to consider our own reaction to any similar Soviet announcement of intent. The very thought that the Soviet Union might plan to deploy effective strategic defense would certainly produce a most energetic American response, and the first and most important element of that response would be a determination to ensure that a sufficient number of our own missiles would always get through.

Administration spokesmen continue to talk as if somehow the prospect of American defensive systems will in and of itself lead the Soviet government to move away from strategic missiles. This is a vain hope. Such a result might indeed be conceivable if Mr. Reagan's original dream were real—if we could somehow ever deploy a *perfect* defense. But in the real world no system will ever be leak-proof; no new system of any sort is in prospect for a decade and only a fragmentary capability for years thereafter; numerous powerful countermeasures are readily available in the meantime, and what is at stake from the Russian standpoint is the deterrent value of their largest and strongest offensive forces.

In this real world it is preposterous to suppose that Star Wars can produce anything but the most determined Soviet effort to make it fruitless. Dr. James Fletcher, chairman of an Administration panel that reviewed the technical prospects after the President's speech, has testified that "the ultimate utility . . . of this system will depend not only on the technology itself, but on the extent to which the Soviet Union agrees to mutual defense arrangements and offense limitations." The plain implication is that the Soviet Union can reduce the "utility" of Star Wars by refusing just such concessions. That is what we would do, and that is what they will do.

Some apologists for Star Wars, although not the President, now defend it on the still more limited ground that it can deny the Soviets a first-strike capability. That is conceivable, in that the indefinite proliferation of systems and countersystems would certainly create fearful uncertainties of all sorts on both sides. But as the Scowcroft Commission correctly concluded, the Soviets have no first-strike capability today, given our survivable forces and the ample existing uncertainties in any surprise attack. We believe there are much better ways than strategic defense to ensure that this situation is maintained. Even a tightly limited and partially effective local defense of missile fields—itself something vastly different from Star Wars—would require radical amendment or repudiation of the ABM Treaty and would create such interacting fears of expanding defenses that we strongly believe it should be avoided.

The President seems aware of the difficulty of making the Soviet Union accept his vision, and he has repeatedly proposed a solution that combines surface plausibility and intrinsic absurdity in a way that tells a lot about what is wrong with Star Wars itself. Mr. Reagan says we should give the Russians the secret of defense, once we find it, in return for their agreement to get rid of nuclear weapons. But the only kind of secret that could be used this way is one that exists only in Mr. Reagan's mind: a single magic formula that would make each side durably invulnerable. In the real world any defensive system will be an imperfect complex of technological and operational capabilities, full understanding of which would at once enable any adversary to improve his own methods of penetration. To share this kind of secret is to destroy its own effectiveness. Mr. Reagan's solution is as unreal as his original dream, and it rests on the same failure of understanding.

There is simply no escape from the reality that Star Wars offers not the promise of greater safety, but the certainty of a large-scale expansion of both offensive and defensive systems on both sides. We are not here examining the dismayed reaction of our allies in Europe, but it is precisely this prospect that they foresee, in addition to the special worries created by their recognition that the Star Wars program as it stands has nothing in it for them. Star Wars, in sum, is a prescription not for ending or limiting the threat of nuclear weapons, but for a competition unlimited in expense, duration and danger.

We have come this way before, following false hopes and finding our danger greater in the upshot. We did it when our government responded to the first Soviet atomic test by a decision to get hydrogen bombs if we could, never stopping to consider in any serious way whether both sides would be better off not to test such a weapon. We did it again, this time in the face of strong and sustained warning, when we were the first to deploy the multiple warheads (MIRVS) that now face us in such excessive numbers on Soviet missiles. Today, 15 years too late, we have a consensus that MIRVS are bad for us, but we are still deploying them, and so are the Russians.

IV

So far we have been addressing the question of new efforts for strategic defense with only marginal attention to their intimate connection with the future of the most important single arms control agreement that we and the Soviet Union share, the Anti-Ballistic Missile Treaty of 1972. The President's program, because of the inevitable Soviet reaction to it, has already had a heavily damaging impact on prospects for any early progress in strategic arms control. It has thrown a wild card into a game already impacted by mutual suspicion and by a search on both sides for unattainable unilateral advantage. It will soon threaten the very existence of the ABM Treaty.

That treaty outlaws any Star Wars defense. Research is permitted, but the development of space-based systems cannot go beyond the laboratory stage without breaking the Treaty. That would be a most fateful step. We strongly agree with the finding of the Scowcroft Commission, in its final report of March, 1984, that "the strategic implications of ballistic missile defense and the criticality of the ABM Treaty to further arms control agreements dictate extreme caution in proceeding to engineering development in this sensitive area."

The ABM Treaty stands at the very center of the effort to limit the strategic arms race by international agreements. It became possible when the two sides recognized that the pursuit of defensive systems would inevitably lead to an expanded competition and to greater insecurity for both. In its underlying meaning, the Treaty is a safeguard less against defense as such than against unbridled competition. The continuing and excessive competition that still exists in offensive weapons would have been even worse without the ABM Treaty, which removed from the calculations of both sides any fear of an early and destabilizing defensive deployment. The consequence over the following decade was profoundly constructive. Neither side attempted a defensive deployment that predictably would have given much more fear to the adversary than comfort to the possessor. The ABM Treaty, in short, reflected a common understanding of exactly the kinds of danger with which Star Wars now confronts the world. To lose the Treaty in pursuit of the Star Wars mirage would be an act of folly.

The defense of the ABM Treaty is thus a first requirement for all who wish to limit the damage done by the Star Wars program. Fortunately the Treaty has wide public support, and the Administration has stated that it plans to do nothing in its five-year program that violates any Treaty clause. Yet by its very existence the Star Wars effort is a threat to the future of the ABM Treaty, and some parts of the announced five-year program raise questions of Treaty compliance. The current program envisions a series of hardware demonstrations, and one of them is described as "an advanced boost-phase detection and tracking system." But the ABM Treaty specifically forbids both the development and the testing of any "space-based" components of an anti-ballistic missile system. We find it hard to see how a boost-phase detection system could be anything but space-based, and we are not impressed by the Administration's claim that such a system is not sufficiently significant to be called "a component."

We make this point not so much to dispute the detailed shape of the current program as to emphasize the strong need for close attention in Congress to the protection of the ABM Treaty. The Treaty has few defenders in the Administration—the President thought it wrong in 1972, and Mr. Weinberger thinks so still. The managers of the program are under more pressure for quick results than for proposals respectful of the Treaty. In this situation a heavy responsibility falls on Congress, which has already shown that it has serious reservations about the President's dream. Interested members of Congress are well placed to ensure that funds are not provided for activities that would violate the Treaty. In meeting this responsibility, and indeed in monitoring the Star Wars program as a whole, Congress can readily get the help of advisers drawn from among the many outstanding experts whose judgment has not been silenced or muted by co-option. Such use of independent counselors is one means of repairing the damage done by the President's unfortunate decision to launch his initiative without the benefit of any serious and unprejudiced scientific assessment.

Cooperating in Outer Space

The Congress should also encourage the Administration toward a new and more vigorous effort to insist on respect for the ABM Treaty by the Soviet government as well. Sweeping charges of Soviet cheating on arms control agreements are clearly overdone. It is deeply unimpressive, for example, to catalogue asserted violations of agreements which we ourselves have refused to ratify. But there is one quite clear instance of large-scale construction that does not appear to be consistent with the ABM Treaty—a large radar in central Siberia near the city of Krasnoyarsk. This radar is not yet in operation, but the weight of technical judgment is that it is designed for the detection of incoming missiles, and the ABM Treaty, in order to forestall effective missile defense systems, forbade the erection of such early warning radars except along the borders of each nation. A single highly vulnerable radar installation is of only marginal importance in relation to any large-scale break-out from the ABM Treaty, but it does raise exactly the kinds of questions of intentional violation which are highly destructive in this country to public confidence in arms control.

On the basis of informed technical advice, we think the most likely purpose of the Krasnoyarsk radar is to give early warning of any attack by submarine-based U.S. missiles on Soviet missile fields. Soviet military men, like some of their counterparts in our own country, appear to believe that the right answer to the threat of surprise attack on missiles is a policy of launch-under-attack, and in that context the Krasnoyarsk radar, which fills an important gap in Soviet warning systems, becomes understandable. Such understanding does not make the radar anything else but a violation of the express language of the Treaty, but it does make it a matter which can be discussed and resolved without any paralyzing fear that it is a clear first signal of massive violations yet to come. Such direct and serious discussion with the Soviets might even allow the two sides to consider together the intrinsic perils in a common policy of launch-under-attack. But no such sensitive discussions will be possible while Star Wars remains a non-negotiable centerpiece of American strategic policy.

Equal in importance to defending the ABM Treaty is preventing hasty overcommitment of financial and scientific resources to totally unproven schemes overflowing with unknowns. The President's men seem determined to encourage an atmosphere of crisis commitment to just such a manner of work, and repeated comparisons to the Manhattan Project of 1942–45, small in size and crystal-clear in purpose by comparison, are not comforting. On the shared basis of conviction that the President's dream is unreal, members of Congress can and should devote themselves with energy to the prevention of the kind of vested interest in very large-scale ongoing expenditures which has so often kept alive other programs that were truly impotent, in terms of their own announced objectives. We believe that there is not much chance that deployments remotely like those currently sketched in the Star Wars program will ever in fact occur. The mere prospect of them will surely provoke the Russians to action, but it is much less likely that paying for them will in the end make sense to the American people. The larger likelihood is that on their way to oblivion these schemes will simply cost us tens and even hundreds of billions of wasted dollars.[4]

In watching over the Star Wars budget the Congress may find it helpful to remember the summary judgment that Senator Arthur Vandenberg used to offer on programs he found wanting: "The end is unattainable, the means hare-brained, and the cost staggering." But at the same time we believe strongly in the continuation of the long-standing policy of maintaining a prudent level of research on the scientific possibilities for defense. Research at a level ample for insurance against some Soviet surprise can be continued at a fraction of the cost of the present Star Wars program. Such a change of course would have the great advantage of preventing what would otherwise be a grave distortion of priorities not only in defense research but in the whole national scientific effort.

V

This has not been a cheerful analysis, or one that we find pleasant to present. If the President makes no major change of course in his second term, we see no alternative to a long, hard, damage-limiting effort by Congress. But we choose to end on a quite different note. We believe that any American president who has won reelection in this nuclear age is bound to ask himself with the greatest seriousness just what he wants to accomplish in his second term. We have no doubt of the deep sincerity of President Reagan's desire for good arms control agreements with the Soviet Union, and we believe his election night assertion that what he wants most in foreign affairs is to reach just such agreements. We are also convinced that if he asks serious and independent advisers what changes in current American policy will help most to make such agreements possible in the next four years, he will learn that it is possible to reach good agreements, or possible to insist on the Star Wars program as it stands, but wholly impossible to do both. At exactly that point, we believe, Mr. Reagan could, should, and possibly would encourage the serious analysis of his negotiating options that did not occur in his first term.

We do not here explore these possibilities in detail. They would certainly include a reaffirmation of the ABM Treaty, and an effort to improve it by broadening its coverage and tightening some of its language. There should also be a further exploration of the possibility of an agreement that would safeguard the peaceful uses of space, uses that have much greater value to us than to the Soviets. We still need and lack a reliable cap on strategic warheads, and while Mr. Reagan has asked too much for too little in the past, he is right to want reductions. He currently has some advisers who fear all forms of arms control, but advisers can be changed. We are not suggesting that the President will change his course lightly. We simply believe that he does truly want real progress on arms control in his second term, and that if he ever comes to understand that he must choose between the two, he will choose the pursuit of agreement over the demands of Star Wars.

We have one final deep and strong belief. We think that if there is to be a real step away from nuclear danger in the next four years, it will have to begin at the level of high politics, with a kind of communication between Moscow and Washington that we have not seen for more than a decade. One of the most unfortunate aspects of the Star Wars initiative is that it was launched without any attempt to discuss it seriously, in advance, with the Soviet government. It represented an explicit expression of the President's belief that we should abandon the shared view of nuclear defense that underlies not only the ABM Treaty but all our later negotiations on strategic weapons. To make a public announcement of a change of this magnitude without any effort to discuss it with the Soviets was to ensure increased Soviet suspicion. This error, too, we have made in earlier decades. If we are now to have renewed hope of arms control, we must sharply elevate our attention to the whole process of communication with Moscow.

Such newly serious communication should begin with frank and explicit recognition by both sides that the problem of nuclear danger is in its basic reality a *common* problem, not just for the two of us, but for all the world—and one that we shall never resolve if we cannot transcend negotiating procedures that give a veto to those in each country who insist on the relentlessly competitive maintenance and enlargement of what are already, on both sides, exorbitantly excessive forces.

If it can ever be understood and accepted, as a starting point for negotiation, that our community of interest in the problem of nuclear danger is greater than all our various competitive concerns put together, there can truly be a renewal of hope, and a new prospect of a shared decision to change course together. Alone among the presidents of the last 12 years, Ronald Reagan has the political strength to lead our country in this direction if he so decides. The renewal of hope cannot be left to await another president without an appeal to the President and his more sober advisers to take a fresh hard look at Star Wars, and then to seek arms control instead.

1. There has been an outpouring of technical comment on this subject, and even in a year and a half the arguments have evolved considerably. Two recent independent analyses on which we have drawn with confidence are *The Reagan Strategic Defense Initiative: A Technical, Political, and Arms Control Assessment*, by Sidney D. Drell, Philip J. Farley and David Holloway, a special report of the Center for International Security and Arms Control, July 1984 (Stanford: Stanford University, 1984); and *The Fallacy of Star Wars* (based on studies conducted by the Union of Concerned Scientists and co-chaired by Richard L. Garwin, Kurt Gottfried, and Henry W. Kendall), John Tirman, ed. (New York: Vintage, 1984).
2. Cited in Sidney Drell *et al.*, *op. cit.*, p. 105.
3. Richard Nixon has analyzed the possible impact of new defensive systems in even more striking terms: "Such systems would be destabilizing if they provided a shield so that you could use the sword." *Los Angeles Times*, July 1, 1984.
4. The Russians have their own program, of course. But they are not about to turn our technological flank in the technologies crucial for ABM systems. "According to the U.S. Department of Defense, the United States has a lead in computers, optics, automated control, electro-optical sensors, propulsion, radar, software, telecommunications, and guidance systems." Drell *et al.*, *op. cit.*, p. 21.

Star Wars requires analysis on the level of politics and strategy, such as Bundy and his colleagues provide; but it also calls for examination on the level of psychology and myth. Co-founder of Global Education Associates, and co-author of *Toward a Human World Order,* Pat Mische argues that "we need to see space not as a new frontier for warfare, but as a new frontier for a sacred journey—a journey into community and our own deepest essence."*

As her title suggests, Mische sees a relation between what we do in outer space and in our own inner space. "Is it possible," she asks, "that our drive to possess, control, and weaponize larger and larger spheres of the planet—and now outer space—is inversely proportionate to our ability to see, inhabit, and master inner space?" To the extent this is so, outer peace depends upon the experience of inner peace. In this sense, apart from its competitive aspect, the Strategic Defense Initiative is a displacement outward of work that needs to be done, not in weapons labs, but on our souls—on the "dead souls" on both sides of the arms race.

Star Wars and the State of Our Souls

Patricia M. Mische

THOUSANDS OF YEARS AGO, OUR TRIBAL ANCESTORS believed that the whole earth—land, water, sky—was inhabited by sacred forces. The whole earth, with its myriad life forms, belonged to the gods. The human task was to learn to bring one's life into harmony with these more powerful, sacred forces. There was a deep belief that the human was not over or separate from the earth, but was part of her life. What one did to the earth one did to oneself.

By the beginning of World War II, the dominant worldview had radically changed. By then the land had been penetrated, militarized,

***Quotes in the introduction, and the following text, are drawn from *Star Wars and the State of Our Souls,* by Patricia M. Mische (Winston Press, 1985).**

and desacralized with tanks, bazookas, mines, and poisonous gas. The waters had been penetrated, militarized, and desacralized with submarines, mines, and torpedoes. And with the bomber even the skies had been penetrated and militarized.

Still, in those "innocent" years, while holocausts were perpetrated on earth, outer space remained sacred space. Buck Rogers lived in comic-book fantasies, but no one had yet gone to the moon. The notion of "star wars" seemed too remote then to generate fears among people on earth. Outer space was no-man's place. Here dwelt Orion, Taurus, Virgo, Aquarius, the Seven Sisters, and other mythical beings, but not mere mortals.

Space was the last place safe from human wars. And it was the last refuge of the image of a sacred presence in the universe. When we prayed we looked heavenward, for the god of creation no longer lived on a militarized earth but had receded in our images to the outer reaches of space. Space was the last dwelling place of the sacred. But soon some began to see space as the new frontier—the high frontier—for militarization. Is it any coincidence, then, that in the late fifties and sixties, when we began to penetrate and then to militarize space, some began to proclaim the death of God altogether?

It has become commonplace to say that we are at a new crossroads in the arms race. Every decade of this century has brought new, more dangerous weapons systems. But we are at more than just another crossroads in the arms race. We are at a new crossroads in our perceptions of national security. We are at a new crossroads in our relationship with outer space. We are at a new crossroads in our relationship with the earth. And we are at a new crossroads in our relationship with the sacred.

Space exploration has opened the way for a new vision of the earth. The modern atomistic worldview in which more than 150 separately acting nation states compete against one another for security stands in conflict with the lessons learned in space: *There is only one air system, one water system, one land system, and one indivisible space system within which all earth societies subsist.*

Security based on territorial boundaries and divisions is appropriate only for those who still believe that the earth is flat and the air divisible. But while many have begun to look to space for new approaches to security, have they been asking the right question? Or have they simply been projecting old worldviews out to the stars?

From Space Militarization to Weaponization?

The *militarization* of space has proceeded over the past decades with little public attention and almost no public protest. According to Representative Joe Moakley (D-MA) the amount the Pentagon expends on its space activities exceeds the entire budget of NASA, the civilian space program.[1] And increasing portions of NASA's budget have been going for purely military purposes, despite the fact that NASA was chartered by Congress to develop the peaceful uses of space. Thus three-fourths of current space expenditures are for military purposes.[2]

But until recently these military purposes were relatively benign and even helpful to peace. They generally were related to communications, navigation, surveillance, and other nonweapon applications that enhanced stability, minimized the possibility of a surprise attack, and made arms control possible by providing national technical means of verification.[3]

Now, however, the militarization of space is moving toward a new, more dangerous phase: *weaponization.* This includes the introduction of weapons that are either (1) directed *at* space objects, whether from bases on land, sea, or air or (2) weapons, for whatever purpose, based *in* space.[4] The weaponization of space would more likely reap war than peace.

The militarization of space is an accompished fact. It will be very hard to change that unless and until disarmament has been largely accomplished. However, the weaponization of space can still be prevented.

Space Dedicated to Peace

Until now the temptation to weaponize space has been suppressed each time it was raised. One reason was the lack of the technical means to do so. Another was the resistance of world public opinion. Early on in space exploration, the competition between the two superpowers, which helped spur technological advances on both sides, was tempered by public pressure to preserve space for peaceful uses.

Cooperating in Outer Space

In the late 1950s the Soviets had put the world on alert with Sputnik. The U.S. responded with a massive mobilization of its educational and scientific institutions. The race for space was on. A few years later, in 1962, John Glenn became the first U.S. astronaut to orbit the earth. In 1969, U.S. astronauts landed on the moon while the world watched via television. The view of the earth rising from the moon's horizon took our breath away. It awakened our imagination. It helped us see the earth in a new way: as one, interdependent, vulnerable planet. It seemed for a time that this achievement, this new vision, would bring all of us on earth closer together.

There were great hopes for the peaceful uses of space. As early as 1959 and through 1962 the Western powers and the USSR had put forward various proposals for banning the use of outer space for military purposes as part of successive plans for general and complete disarmament, but agreement was never reached.

In 1966 the General Assembly of the United Nations approved the Outer Space Treaty[5] after reconciling drafts submitted by both the United States and the Soviet Union. This treaty affirmed "the common interest of all [hu]mankind in the . . . exploration and use of outer space for peaceful purposes." Among other things, signatory states pledged:

that "the exploration and use of outer space . . . shall be carried out for the benefit and in the interests of all countries, irrespective of their degree of economic or scientific development, and shall be the province of all [hu]mankind";

that the exploration and use of outer spce would be guided by "the principle of cooperation and mutual assistance";

that any State party to the treaty that launched objects into outer space, or from whose territory objects were launched, would accept international liability for any damage to another State party that might result;

that they would "not place in orbit around the earth any nuclear weapons or any other kinds of weapons of mass destruction"; and

that they would not establish any military bases, installations and fortifications, or test any type of weapon or conduct military maneuvers on celestial bodies.

The treaty was opened for signature at Washington, London, and Moscow on January 27, 1967. On April 25 the U.S. Senate gave unanimous consent to its ratification. It entered into force on October 10, 1967. By 1982, eighty-nine nations had signed, ratified, or acceded to the treaty.

With this treaty in place the world's people felt relatively secure about the peaceful uses of space. In the years since its adoption it has contributed to the peaceful space environment that made U.S. and Soviet cooperation in space possible for a time and that enabled the development of world communication systems to bring all of us around the world closer together.

Thus, until recently, "Star Wars" seemed to be something that would remain the subject only of science fiction.

Preparing for Star Wars

Preparation for Star Wars did not begin suddenly with President Reagan. While Reagan focused media attention on it in his famous Star Wars speech of March 23, 1983, suggesting that the U.S. should pursue further research and development in this area, the research has, in fact, been underway for many years.

My first awareness that space weapons had moved out of the realm of science fiction and onto the military drawing boards came in 1979 during President Carter's administration. At the time the U.S. Department of State was trying to arouse public support for SALT II and was conducting a series of citizen briefings. I had just come from one of these and was enroute from Washington, D.C., to St. Louis, where I was going to give a talk. Next to me on the plane was an Army major who taught weapons systems for the Army. He had just come from his monthly Pentagon briefings. There was a blizzard below and the plane had to circle for several hours. Free drinks were served and camaraderie flourished.

I asked the major about the security implications of SALT II. "If you ask me," he said, "SALT doesn't affect our security one way or another. We already have more nuclear weapons than we need, so putting a ceiling on the numbers doesn't have any effect on U.S. security. Besides," he continued, "I'm already teaching about new weapons not even included in the SALT negotiations: lasers, particle beams, killer satellites. . . . "

According to the Institute for Space and Security Studies: "The idea of space-based ABM systems goes back about a quarter of a century and is as old as ballistic missiles themselves. For the last decade [the U.S. has been] spending at least a hundred million dollars a year on the development of such weapons—an amount that has increased to about a billion a year."[6]

A number of types of space weapons have been undergoing research including anti-satellite weapons (commonly referred to as ASATs and aimed at destroying an opponent's satellites) and various space-related anti-ballistic missile (ABM) or ballistic-missile defense (BMD) systems (commonly referred to as Star Wars weapons and aimed at destroying an opponent's launched ballistic missiles before they reach their target). The Star Wars ABM or BMD systems include high-energy lasers, particle beams, microwaves, orbiting mirrors (to be used with lasers), rocket-powered kill vehicles and high-velocity buckshot. Different types of high-energy laser systems are being explored, including chemical-powered lasers and nuclear-powered X-ray lasers. The "High Frontier" proposal that is being vigorously promoted by the Heritage Foundation is a multi-tiered plan that would utilize a number of these systems. It involves both space-borne and ground systems to "filter" missiles at different stages in their trajectory toward a target.

Proponents Argue Security, Morality, and Soviet Ruin

Depending on their audience, proponents of these space weapons make different claims. When appealing to currently strong anti-nuclear sentiment they argue, like President Reagan, that space weapons will render nuclear weapons "impotent and obsolete,"[7] or, like Lt. General Daniel Graham (Ret.), that a space defense will "remove the brooding menace of nuclear war and end the immorality of Mutual Assured Destruction."[8] In his Star Wars speech, President Reagan told Americans that their hope for the future depended on a system of space weapons to shoot down enemy missiles. He appealed to survival instincts and strongly implied that the weapons would protect people from a nuclear missile attack.

This author agrees with President Reagan's stated goal to bring an end to the threat of nuclear destruction. But as we will see, the problem with Star Wars systems is precisely that they would not end the risk of nuclear war; they would increase that danger.

Many proponents of space weapons openly admit space weapons would not be a foolproof protection against a nuclear attack. They advocate space weapons for reasons quite different from those articulated by the president or General Graham and do not even pretend to want the elimination of nuclear weapons. Here are some of their reasons.

Undercut the Nuclear Freeze Movement

For example, Greg Fossedal, a special consultant to the Heritage Foundation's High Frontier project, writing in the *Conservative Digest*, argued that proposals for a space defense would give the administration "an opportunity to . . . fast-thaw the nuclear-freeze movement." To make clear what he had in mind, he elaborated: "Armed with the detailed heritage study . . . President Reagan would have the secret weapons needed to undercut the nuclear freeze crusade."[9]

It is hard to see how a space defense would undercut the nuclear-freeze movement if its true objective was to eliminate nuclear weapons. Few freeze proponents would argue against that. Such anti-freeze posturing only makes it clear that space weapons do not have as their true purpose the elimination of nuclear weapons.

Observers of the Star Wars debate should not be fooled by proponents who argue that space weapons will eliminate nuclear weapons. In a secret plan entitled "BMD and Arms Control" the Heritage Foundation advises space-weapons proponents to usurp the language and cause of the anti-nuclear movement to advance space weapons. A few quotes from this document indicate how this fits into an overall strategy:

"Keep BMD program alive in 1984 and make it impossible to turn off by 1989" . . . "Permit U.S. to move ahead forcefully and *unilaterally*" (emphasis in original) . . . "represented as a *bilateral* effort—one with Soviet reciprocation and participation" . . . "Make it politically risky for BMD opponents to invoke alleged 'arms control arguments' against an early-IOC or any other BMD system" . . . "Disarm

BMD opponents by stealing their language and cause . . . their explicit or de facto advocacy of classical anti-population war-crimes" . . . "With appropriate political and emotional packaging, this approach may be able to tap the freeze constituency (i.e., the 'do *something*' approach to arms control."[10]

Reduce Vulnerability, Preserve Technological Lead, Limit Damage

Space weapons are also being proposed by Fossedal and others as a means to "slam shut the window of U.S. vulnerability, . . . preserve America's narrowing technological lead in space, [and] greatly limit the damage from any Soviet first-strike attack."[11]

Reducing vulnerability and preserving a technological lead are quite different from eliminating nuclear weapons. They could lead toward a build-*up*, not a build-down. It is generally acknowledged that land-based missiles are vulnerable. However, this is more than compensated for by less vulnerable submarine and air-based missiles which could wipe out every Soviet city in a retaliatory strike. But if reducing the vulnerability of land-based missiles still seems necessary despite the destructive power of the other systems, then there are more reliable and far less costly ways to do it than a new arms race in space.

And if a technological lead in space is desired, why not lead in technology for peaceful space development? The Soviets are currently ahead of the U.S. in some areas of non-military space development. If competition is the game being sought, why not try to match and excel the Soviets in scientific, medical, industrial, and humanitarian space programs? At least this competition would provide some real benefits to the nation and the world.

Finally, "limiting the damage" of a nuclear attack is a much more "modest" goal than "rendering nuclear weapons impotent and obsolete." If this means reducing the numbers killed to "only" a few million instead of hundreds of millions and the numbers of cities destroyed from hundreds to a handful, some think it would be worth the great expenditure. But if the goal truly is to save people and cities, then the best way to do it is to negotiate a comprehensive ban on all nuclear weapons. Let *no* people be killed and *no* cities destroyed in a nuclear war. Merely "limiting the damage" is so modest an aim it is not worth pursuing. Who could offer up their children or their cities to be among the victims so that their country could engage in a "safer" nuclear war?

Confound the Soviets

Some proponents seem primarily motivated by anti-Soviet sentiments and a desire to avoid detente and arms-control negotiations. They argue that space weapons would make detente with the Soviet Union unnecessary.[12] This argument is based on the theory that if we can shoot down anything the Soviets send our way, there is no need to establish better relations or to negotiate arms reductions. This is truly wishful thinking. Few, if any, experts in this field are prepared to rest their reputation or the nation's security on the reliability of these systems for shooting down all the thousands of missiles the Soviet Union could launch.

Some go even further in being guided by anti-Soviet feelings. They don't want to avoid the Soviets; they want to confound them. Edward Teller, who worked with Oppenheimer on the A-bomb and has defended nuclear weapons research and development ever since, now wants the U.S. to use small nuclear bombs to power space-based laser weapons. He condemns critics in the scientific community who find fault with Star Wars research. "One cannot work on military research without a positive attitude," he says. But when asked what such research would accomplish, he couldn't think of anything more promising to say than that it would force the Russians to increase their expenditures greatly. "If that happens, we would have accomplished something."[13] (Teller said nothing about what it would do to the U.S. economy or whether, in trying to drive the Soviets to economic ruin, the U.S. might not be bringing economic ruin upon itself.)

Sheer anti-Sovietism takes other forms. Some believe that all Americans who love their country should be for space weapons simply, and even if only, because the Soviet Union is against them. They reason that if the Soviet Union is *against* something, there must be something *good* in it for the United States. But by the same logic, if the Soviets are against child abuse, the U.S. should be for it. If the Soviets are for health and survival, the U.S. should be against it.

Power, Domination, and the Thrills of War

There are still other motivations for space weapons: power, domination, and the romantic notion that fighting a state-of-the-art war in space could be an exhilarating event.

Lt. General Richard C. Henry, Deputy Commander of Air Force Space Command (Ret.) described space as "the new high ground of battle."[14] Edward C. Aldridge, Under-Secretary of the Air Force, prophesied, "We do not have to stretch our imagination very far to see that the nation that controls space may control the world."[15] And General Robert T. Marsh, Commander, Air Force Systems Command, pointed to the exciting opportunities for new types of warfare: "We should move into war-fighting capabilities—that is ground-to-space war-fighting capabilities, space-to-space, space-to-ground."[16]

Opponents Argue Cost, Ineffectiveness, Technical Obstacles, and Danger of War

Opposing the weaponization of space are many military, scientific, and arms-control experts and organizations such as the Union of Concerned Scientists, the Center for Defense Information, the Institute for Space and Security Studies, and the Institute for Security and Cooperation in Outer Space. Prominent individuals opposing space weapons include Dr. Hans Bethe, a Nobel laureate and physicist who worked on the Manhattan project; Drs. George Rathjens, Kosta Tsipis, and Jack Ruina of M.I.T.; Drs. Sidney Drell and Wolfgang Panofsky, Stanford physicists; Dr. John Steinbrunner, head of international relations at the Brookings Institution in Washington; Richard L. Garwin of Harvard; and many others.

The reasons for their opposition vary from practical, technical problems to cost factors and strategic security considerations. Directly contradicting and contesting the claims of the space weapons advocates, they present the following arguments.

High Cost

Space weapons would cost incredible sums of money. The Reagan administration has asked for $33 billion for research alone for the six fiscal years ending September 30, 1991. Military analysts estimate it would cost between $70 billion and $100 billion by 1993 to develop and test the technology.[17] But that does not yet include the cost of actually producing and deploying the systems. According to one Pentagon estimate, by the time the systems are in place the bill could rise to $500 billion.[18] Others, including former Secretary of Defense James R. Schlesinger, put the total cost at a trillion dollars.[19] This price would be overwhelming at any time. It is especially hard to justify at a time of grave federal deficit and major cutbacks in social services and human needs.

Civilian Space Programs Undermined

The development of space weapons would undermine peaceful uses of space by taking money, talent, and resources away from new programs that could benefit humanity, and by threatening existing civilian communications, mapping, weather monitoring, search-and-rescue, and other programs.

Technical Obstacles

Richard D. DeLauer, U.S. Undersecretary of Defense for research, acknowledged in Congressional testimony that directed-energy weapons posed eight serious technical problems, *each* of which would involve "staggering costs" to remedy and require a mobilization of science and engineers as great as or greater than that required to land a man on the moon.[20]

Death of Arms Control

Most space weapons systems would violate existing arms-control treaties. Furthermore, they would threaten the satellite verification systems upon which future and existing arms-control treaties depend. Undermining arms control does not lead to improved national security; it leads to greater *insecurity* by taking the lid off the arms race, by further straining U.S./Soviet relations, and by further weakening an already fragile system of international law. The result could be near anarchy in international relations, with the rule of force, not the rule of law, prevailing.

Vulnerable and Ineffective Defense

Space systems would not provide an effective defense. All the various space-weapons proposals have serious vulnerabilities. For example, says Robert Bowman, president of the Institute for Space and Security Studies, "none of them could survive a single nuclear explosion in space. The radiation and electromagnetic pulse from a single blast could wipe out the sensitive electronic brains of hundreds of battle stations, clearing the way for a following wave of ICBMs."[21]

Easy Countermeasures

Furthermore, any defensive capabilities that might be perfected could be effectively bypassed by countermeasures. And the countermeasures would be far less costly than a space defense.

Already a secret Air Force project called the Advanced Strategic Missile Systems program is working to design and test "penetration aids" to help American missiles sneak past any space defense the Soviets might develop. Included are zigzagging warheads, advanced decoys (to make a defense system contend with a "baffling multiplicity of targets"), clouds of wire bits called chaff, light-reflecting aerosols to confuse sensors, and "defense suppression systems" that home in on and destroy radars to clear the way for nuclear missiles. As a *New York Times* article noted, while the U.S. Defense Department is researching ways to render Soviet missiles impotent, it is simultaneously "trying to assure that American nuclear missiles never meet the same fate."[22]

But if the United States can think up such countermeasures, the Soviets are smart enough to think of them too. Every new weapons system developed by the United States has eventually been matched or countered by the Soviets. Besides measures such as those being tested in the above Air Force program, the Soviets could also "go to depressed-trajectory systems which travel beneath the systems's coverage, shift to cruise missiles, which are impervious to any space-based defense, or merely multiply the number of offensive missiles so as to overwhelm the defense."[23]

With such a variety of easy countermeasures it is not nuclear weapons that will be rendered impotent and obsolete, but the space defenses themselves. Then why waste billions of dollars developing such systems in the first place?

Perpetuation of MAD Policies

Space weapons would not bring an end to the threat of Mutual Assured Destruction (MAD), but are more likely to perpetuate it. Many of the proposed space defense systems, such as General Graham's High Frontier, in actuality are not aimed at protecting *people* in the case of a nuclear attack, but rather at protecting U.S nuclear missile silos! The only purpose for this can be to preserve the capacity for nuclear retaliation. Thus the rationale that these systems will bring an end to policies of nuclear deterrence and nuclear-war fighting is simply not true.

It should not be ignored that the very people who argue for Star Wars systems by claiming these weapons do not kill people are the same ones who invariably support the MX, Pershing II, Trident II, and other offensive weapons, all of which *do* kill people.[24]

First-Strike Potential

While many experts are skeptical about the defensive reliability of the proposed space systems, they point out their potential for offensive use as part of a first strike. The first step would be an anti-satellite (ASAT) attack on a nation's strategic satellite communications, verification, navigation and other information systems, followed by an attack on strategic missile silos to prevent nuclear retaliation. Then the offending nation would use its Star Wars ABM systems to wipe out any remaining missiles launched in retaliation.

The first-strike scenario is not far-fetched. It is commonly acknowledged even by proponents of space weapons that these systems could not handle large numbers of missiles at one time and that, in the event of a massive first strike, large numbers of nuclear missiles would get through the space "defenses." However, space weapons would be more effective against a small number of missiles—for example, the small number remaining after a massive first strike. As the ISSS has noted, "If this is so, such a system would be of little protection against an enemy first strike, but might be very useful in mopping up the few retaliatory missiles remaining after a first strike by the side possessing the defensive system."[25]

Thus nuclear deterrence would not be rendered *unnecessary*, as General Graham has stated, but *inoperative*. The nation with space weapons in place first might not be deterred from a nuclear attack. Rather than preventing a first strike, space weapons could tempt one because they would give a decided "advantage" to the nation striking first.

We must not be fooled by those who claim these systems are for defense only. Proponents of the MX missile claim it is a defensive system because it is designed to destroy an opponent's missiles in their silos. But destroying missiles still in their silos constitutes a first strike. Is a first strike really a defensive act? Or is the word "defense" here merely a euphemism for aggression?

Nuclear War by Computer Error

Finally, *the hair-trigger responses* of these systems make them especially dangerous. Each side's systems could destroy the other's systems in a fraction of a second. As an ISSS report warns, hair-trigger response would take on a whole new meaning when engagements take place at the speed of light. Not only would there be an enormous temptation to shoot first in order to leave the opponent defenseless, but a false indication of an attack against one side's ABMs would necessarily trigger an automatic and immediate response that would lead to a full-scale nuclear exchange.

Concern about hair-trigger, launch-on-warning space systems is not based on abstract speculation. The Reagan administration specified to industry contractors at a November 16, 1984 conference in Pentagon City that it wanted an orbiting supercomputer with artificial intelligence software that would give it "automated rules of engagement and firing doctrine" for its SDI (Star Wars) program. And the industry representatives were told *it would have to be effective within one second of detection of the first missile.*[26]

This means there would be no time for human imput to check for computer error. Yet there have been hundreds of false nuclear alerts over recent years, many of them due to computer error. In those instances it was human input that stopped the missiles and planes and prevented a nuclear war. A one-second trigger time would not even allow the president the prerogative of making the critical decision. The future of the nation and of the whole world would be in the hands of a microchip. And forget about using the hotline to Moscow to check on what is happening: one second is not enough time to get to the phone.

Neither the U.S. nor the USSR wants to destroy itself and civilization in a nuclear war. But a nuclear war by accident is perhaps our greatest risk today, and that risk would be increased a thousand-fold with one-second, launch-on-warning, computer-controlled systems.

Weapons Determine Policy: From War Prevention to Nuclear-War Fighting

In the past it was central to U.S. military philosophy to subject weaponry to strategy. The primary strategy was *prevention* of nuclear war.

Adherence to this policy led to the successful conclusion of the ABM Treaty in 1972. Both superpowers recognized that anti-ballistic missiles would be very costly and that they might not work. Even more importantly, they recognized that ABMs would fuel an *offensive* arms race that would increase the risk of nuclear war.[27]

Up to now military-related space systems have also been subject to a strategy of war prevention. They have been, says Bowman, "strictly peacetime systems, designed to support a strategy of deterrence and not survivable in a conflict situation."[28]

But in the last few years this has been changing: now *weapons have begun to determine strategy*. According to Bowman,[29] MIRVs (Multiple Independently-targetable Reentry Vehicles) theoretically made a first strike advantageous because the multiple warheads on one missile could simultaneously strike a number of missile silos in different locations and thus minimize the possibility of retaliation. This undercut the policy of deterrence and began a shift in strategy toward fighting rather than preventing a nuclear war because it offered an advantage to the nation that struck first.

The U.S. had MIRVs first. Seduced by the lure of technological superiority, the U.S. refused to include a ban on MIRVs in the 1972 SALT I Treaty. But in the absence of such a ban, it took only a few years for the Soviets to develop their own MIRVs. With its own missile silos now in jeopardy, the U.S. was less secure than ever.

U.S. military leaders then announced a "window of vulnerability." The MX (Missile Experimental) was dreamed up to resolve the problem. The idea was to produce a missile system that could survive a first strike, be used in retaliation, and thus maintain a deterrent. But the "survivability" element proved too costly and difficult to achieve. Instead of abandoning the project, however, proponents of the system were so fascinated by other technological innovations that the MX was developed anyway, with a highly accurate guidance system that gave it a first-strike or "silo-busting" capability.

"The MX was a misfit in our deterrent strategy," says Bowman. But instead of changing the weapons to fit the strategy, we changed "our strategy . . . to fit our weapons."[30]

In other words, *commitment to the new first-strike weapons has led to a strategy shift from preventing nuclear war to preparations for fighting one. This shift is now threatening to alter space policy*. Until now, war between the superpowers has been avoided "largely because of the stabilizing influence of space systems," says Bowman. "The military surveillance systems of the United States and the Soviet Union have until now contributed immeasurably to peace by denying the element of surprise to an attacker and thus diminishing the temptation to launch a first strike. By giving each side the knowledge that they could not be taken by surprise, they have reduced the pressures for pre-emptive strikes and led to a considerable lessening of tension. Space systems provide time for analysis, confirmation, consultation, and deliberation, and have made hair-trigger responses unnecessary."[31]

But once they adopted a policy of nuclear-war fighting, U.S. military planners were confronted with a critical choice in space policy. Existing space systems had not been designed for fighting a war, but for preventing war. The choice was either to abandon the MX and other first-strike weapons and return to a policy of war-prevention, or to weaponize space for the eventuality of nuclear-war fighting. In proposing that the country spend hundreds of billions of dollars to produce MX missiles and to develop space weapons, President Reagan indicated his clear choice: to prepare space—and the United States—for *nuclear-war fighting.*[32]

How Will We Choose?

But this choice need not be the choice of the nation. The final decision is not in yet. The president still has time to reevaluate this dangerous direction and lead the world toward peace in outer space. In a certain sense the president of the United States is the president of the world. Although elected by the people of only one nation, the president's choices affect everyone on earth.

Most importantly, in our democracy, it is we the people who will also choose. We cannot afford to abrogate or relinquish our right and freedom to choose on this issue. Whatever the ultimate choice about war or peace in outer space, we who live in this time will, by our decisions and actions, share in either the blame or the credit for the fate of outer space and with it, the fate of the earth and of ourselves.

For the fate of outer space is not something that will happen in isolation, with no bearing on the planet and those who live in her life. The planet dwells within space like a developing child in the womb, and we similarly dwell within the planet's life processes. While worldviews may have changed and ancient wisdoms been forgotten, the vision of our tribal ancestors nevertheless remains as true today as it was in the past: What we do to the earth we do to ourselves. By the same wisdom, what we choose to do in outer space we will do to ourselves.

Our youngest child was born the year astronauts planted an American flag on the moon. A few years later we were invaded on Halloween by hundreds of kids dressed as Luke Skywalker, Princess Leia, Darth Vader, and R2-D2. By the time the first *Star Wars* film took possession of the public imagination, the penetration and militarization of space had become an accomplished fact that bred its own fantasy about a future time when planets would be zapped out at the speed of light.

The administration has repeatedly tried to dissociate the Strategic Defense Initiative and space weaponization from the public identification of it with *Star Wars*. The concern is that the *Star Wars* label belittles this "defense" proposal as mere fantasy that does not merit serious public consideration. Nevertheless, the press and the public persist in using the term. Perhaps this is due to more than mere cynicism about the value of the costly, unproven, and doubtful proposal for a "space defense." The public association with the *Star Wars* film by the same title may have more to do with some of the deeper, unconscious, and frightening implications of the film.

The hidden assumption of *Star Wars*—filmed as it was from the perspective of the far future—was that long ago, in the barbaric, dark ages of the twentieth century, arms control and disarmament efforts had failed. War had been taken beyond earth's oceans, land, and air and had been transported into outer space. En route it had gotten bigger and "better" and more remotely controlled so that even when a whole planet was "eliminated," no one had to hear the screams or see the blood and gore. War was depicted not so much as a human tragedy as a full-screen, live-action video game. There was only a momentary pause and pale regret for the lost planet before the action of *Star Wars* continued to its exciting climax of intergalactic warfare.

The film's unquestioning stance left me saddened. Its failure to raise to a conscious level the question about whether the history of the earth and space could not have gone in another direction if wiser choices had been made back in the twentieth century testified in some way that the earth was already lost. In unquestioningly accepting and mentally preparing for "star wars" we were giving up on the earth and on ourselves and on our children and great-great grandchildren.

We were giving up, too, on our own souls. For ultimately what we do in *outer* space will be a mirror image of our *inner* space, of the state of our souls. It will reflect whether we have learned to name and tame our capacity for violence and destructiveness and to enlarge our capacity for community, cooperation, healing, creating, and loving. It will reflect whether we have succumbed to a Faustian fascination with the technology of destruction or whether we have learned to develop life-enhancing technology to accompany an enlargement of our spirit.

Cooperating in Outer Space

The succeeding two films in the *Star Wars* trilogy began to touch more on this relationship between inner space and outer space. Luke wants to be a Jedi. He must first learn to develop and direct his inner powers as a force for good in the universe. Under the guidance of Yoda, his wise old teacher, he retreats inward to develop self-knowledge and spiritual mastery, and to gain insight, skills, and strength for his greatest test on the path to maturity: his face-to-face encounter with his polar opposite, Darth Vader.

While in training Luke is sent by Yoda on a frightening journey where unknown pitfalls and dangers lurk in dark and hidden places. It is the most important and difficult part of his training. It is a journey into his own soul. As Luke sets off, he asks Yoda, "What is there?" Yoda answers, "Only what you take in there."

While in this space Luke encounters the image of the one he most dreads: Darth Vader. They enter into deadly battle, and in the struggle Luke fells his archenemy. Then he stoops to lift Vader's visor so he can see the face of his enemy. He falls back aghast, filled with terror. The face he sees is his own.

This drama of the unconscious foreshadows the terrible truth Luke is to learn in a more conscious way in outer space when he next encounters Vader. While engaged in deadly combat with his archenemy, Luke is stunned by the revelation that Vader and he are father and son.

For Luke this news is traumatic and unbearable. He cannot accept his relatedness to his enemy. Later Luke is to learn an even deeper and more painful truth. It is more than blood that ties them. They are also connected in their common capacities for love and hate. They who are polar opposites abide in each other. Luke discovers that what he despises and is trying to destroy in Darth Vader exists also in himself. And the heroic qualities and capacity for good and light which he has tried to develop in himself once shone in Darth Vader. Luke, in his crusade to conquer the dark force of hatred and violence, evokes those very characteristics in himself. He begins to employ violence and hatred to defeat his enemy. Seeing this, Darth Vader predicts that the forces of darkness will triumph in the universe.

The warning Yoda gave Luke earlier in his training now comes to light: "Beware of the dark side. Once you start down the dark side you cannot turn back. It takes over." This is the path Vader had chosen before him. Now Luke faces the same test of soul. Yoda had taught, "If you choose the quick and easy path, you will become the agent of evil." He had also taught Luke about the power and the path of the life force. "It makes things grow. Its energy surrounds us and binds us everywhere."

The climax of the trilogy brings a redemptive healing for both father and son. Accepting their intimate connectedness to each other, they reach out in love and compassion. This moment is the gateway to wholeness and peace of soul for both. It is also the gateway to external peace—in this case, peace in intergalactic relations. But for many in the viewing audience, the subtle relationship between the resolution of the inner conflicts of the soul and the conflicts in outer space went largely unnoticed.

Until that moment the central question driving both toward a destructive solution was, Who will win over the other? When they understood and acted in the knowledge of their relatedness, the question changed (although too late for Darth Vader). It became instead, How can we save each other? Changing the question changed everything.

Is it possible that the two superpowers are like Luke Skywalker and Darth Vader? Each functions within a dualistic belief system in which they see the world as divided between the forces of good and the forces of evil. Each believes it is the force for good and the other the force for evil. Each believes it can achieve its own security only at the expense of the other. But in a world of interdependent nations this basic assumption is bound to lead to strategies in which no one will be secure. Like Luke Skywalker and Darth Vader, the two superpowers need to accept at last their intimate relatedness. Else they will forever threaten mutual destruction.

Is it also possible that the way in which the two superpowers choose to respond to the challenge presented by outer space may lead either to their ultimate self-destruction or to their redemption? The capacities for both good and evil dwell in each of us, and we can choose whether and how we will use those capacities to shape history. In deciding the fate of outer space, we can choose to allow the triumph of violence in our psyches and souls, imposing our darkness on the earth and on the universe. Or we can choose to recognize and act on the knowledge of our indwelling in one another and in the earth. We can choose to utilize our creative capacities and technological genius for leading the human community in the cooperative development of outer space for the advancement of peace and human well-being on earth.

We are at a critical point in the evolution of the earth and in our development as human beings. As human beings we are from the earth and of the earth. But we are the earth in a new phase of its evolution. With our new knowledge and science we can intervene in the earth's life-sustaining capacities. We can now determine the next stages in the earth's evolution.

We are also approaching that point in the earth's and our own development when we will be able to go beyond being earthbound creatures. We are standing at the gateway to the heavens, able to choose whether we should enter that gateway with the weapons of war or with the tools for peace.

Questions at the Gateway to the Heavens

Yes, we are standing at the gateway to the heavens. But are we asking the right questions?

We should be asking questions such as "What can outer space teach us about who we are, our relatedness to each other, and how to live together on one earth?" or "What can it teach us about the cosmos, about the solar system, about our planet as a subsystem within the larger cosmo-system, about the origins of the universe and its ongoing genesis? Can we learn something in space about what it means to be human? to be a human community? to be planetary creatures? Can our sojourn into outer space help us learn how to resolve some of our critical problems on earth?"

Instead, our leaders are poising to take us into battle competition for possession and control of yet another global commons. We have chosen to struggle for identity, power, and mastery by making greater and greater external territorial claims, but we have ignored the inner territory of our minds and souls and the larger territory of the universe that we can never possess and are only on the fringes of being able to explore or understand.

Is it possible that our drive to possess, control, and weaponize larger and larger spheres of the planet—and now outer space—is inversely proportionate to our ability to see, inhabit, and master our inner space? Is it possible that when we fail to identify and acknowledge the existence of our own dark side, we project it outward onto others and now into outer space? Is it possible that when we fail to see and own our dark side, we also limit our capacity to see and know ourselves as children of the light? to fully know ourselves as "blessed" and "graced" and able to be much more than we have yet become?

We can take pride in the many technological achievements of this century. These new technologies offer great promise for future life on earth and pave the way for our voyage to the stars. But they also bring us to a critical point in our budding planetary civilization. A civilization must not destroy itself once it has mastered the technology to do so. We are at that precise point of danger. We have developed the technology that can take us into outer space, but we have not yet developed with it the moral maturity, wisdom, or spiritual depth—nor even the language—to think about and manage our new technological powers in ways that will ensure our future survival and well-being on earth and open us to the fullness of the heavens and the fullness of ourselves. Our greatest challenge is to overcome this tragic lag.

The Banality of Evil

At the end of World War II, the philosopher Hannah Arendt was an observer at the trials of Nazi war criminals. Later she wrote about the banality of evil.

She had been surprised to see how ordinary those who committed atrocities looked. They looked like other ordinary human beings who loved their children, their dogs, fine music, and art. Those ordinary people had extraordinary powers over life and death, but their tragic flaw was that they had no mental or moral tools to think about what they were doing. So they marched in blind obedience to the dictates of their political leader or their technological obsessions. They were guided not by the effects of their actions and technology on human beings, but by technological and logistical problem-solving as an end in itself: how to make the technology of gas chambers and crematoria more effective and efficient.

In the prologue to her book *The Human Condition*, Arendt reflected further on the dangers of the lag between technological development and our ability to think about what we are doing:

"The trouble concerns the fact that the 'truths' of the modern scientific world view, though they can be demonstrated in mathematical formulas and proved technologically, will no longer lend themselves to normal expression in speech and thought. . . . We do not know whether this situation is final. But it could be that we, who are earth-bound creatures and have begun to act as though we were dwellers of the universe, will forever be unable to understand, that is to think and speak about the things which nevertheless we are able to do. In this case, it would be as though our brain, which constitutes the physical, material condition of our thoughts, were unable to follow what we do, so that from now on we would indeed need artificial machines to do our thinking and speaking. If it should turn out to be true that knowledge (in the modern sense of know-how) and thought have parted company for good, then we would indeed become the helpless slaves, not so much of our machines as of our know-how, thoughtless creatures at the mercy of every gadget which is technologically possible, no matter how murderous it is."[33]

Lest we think that these words, written in 1958, were directed only at events such as Hitlerian genocides or atomic bombs, and have no bearing on current proposals to develop a space "defense," let us be mindful that Arendt begins her book be referring to the 1957 launching into space of the first man-made satellite: "This event, second in importance to no other, not even to the splitting of the atom, would have been greeted with unmitigated joy if it had not been for the uncomfortable military and political circumstances attending it."[34]

Arendt is not down on science and technology. She rejoices in their triumphs, real and potential. Rather, she is concerned about our lack of a language to think about what we are doing, and our inability to think, discuss, and make moral judgments about the applications of technology—an inability that results from our lack of language. What she mourns is the *reversal of contemplation and action*: that we "do" first and contemplate the consequences of our actions only *after* the "doing," or not at all:

"The sciences today have been forced to adopt a 'language' of mathematical symbols which, though it was originally meant only as an abbreviation for spoken statements, now contains statements that in no way can be translated back into speech. The reason why it may be wise to distrust the political judgment of scientists *qua* scientists is not primarily their lack of 'character'—that they did not refuse to develop atomic weapons—or their naivete—that they did not understand that once these weapons were developed they would be the last to be consulted about their use—but precisely the fact that *they move in a world where speech has lost its power.* And whatever men do or know or experience can make sense only to the extent that it can be spoken about" (emphasis added).[35]

A *world where speech has lost its power*: Hundreds of thousands of people are homeless, and lack adequate nutrition, education, employment, and health care; farmers are losing their farms. Yet the administration that wants to cut assistance to them to ease the federal deficit is the same government asking that the nation submit to billions of dollars for new weapons that in the end will not save people, farms, or cities, and certainly not the budget, but only nuclear-missile silos.

A *world where speech has lost its power*: A TV ad shows a missile striking another missile in flight. The ad solicits young men for a Navy career. The ad reminds one of an Atari game. It appeals to the young video addict who wants to play the game for real, in bigger, better ways. It is *Star Wars* moved from video and movie fantasies into mindless, wordless acceptance. The ad gives a subliminal impression that a space defense would work. It does not show how often the tests have failed.

A *world where speech has lost its power*: The administration holds a series of meetings for industry representatives to generate interest in Star Wars-related contracts. Using the language of technological seduction, a promotional brochure entitled "Finding Your Role in Strategic Defense Initiative Battle Management" lures industry readers to get their share of the billions of dollars budgeted by the Department of Defense for Star Wars research. It tempts them with a cutting-edge opportunity: "Will you find a place in this state of the art cornucopia—expected to be a greater undertaking than the Apollo program?"[36]

Thus in the world of Star Wars planning, speech is being twisted to tempt persons to sell their souls and minds for a lucrative contract and the momentary exhilaration of being on "the inside" of the latest technology game in town.

Arendt writes: "There may be truths beyond speech, and they may be of great relevance to man in the singular, that is, to man in so far as he is not a political being, whatever else he may be. Men in the plural, that is men in so far as they live and move and act in this world, can experience meaningfulness only because they can talk with and make sense to each other and to themselves."[37]

Now, more than ever, we need to heed the words of Arendt: "*What I propose is very simple: it is nothing more than to think what we are doing.*" The banality of the evil that may be perpetrated on the planet and on ourselves in taking the arms race into space is that we may allow it to be done to us and in our name, not because it meets any real security need, but because those political and technological elites whom we have chosen or allowed to make decisions in the name of our security cannot "think what they are doing."

And we may allow it because we too, like the good Germans, "cannot think what we are doing." As a consequence we may strike a Faustian bargain in which we sell the future of our children and grandchildren, along with true national and world security, for the sake of indulging our obsession with new weapons technologies and in blind obedience to doing whatever appears technologically possible.

Who Owns the Earth and the Stars?

Throughout human history societies have claimed possession of increasingly larger territories—from tribal lands to city-states and kingdoms to national territories. It was symptomatic of the European "Age of Discovery" (when it was learned to everyone's surprise that the world was not flat and that just around the world from them—in the neighborhood, so to speak—were other lands, other waters, other human societies) that nations began to claim other parts of the planet (whole continents sometimes, along with all the inhabitants) as their national possession.

Today the superpowers claim "national spheres of interest" and send troops around the world to "protect" *their* spheres. Now, as we go into outer space, our national egos and will to power, along with our eyes for territorial expansion, have grown bigger than the planet.

Nations used to defend their borders from *within* those national borders. But gradually they moved their weapons into the global commons—the spheres that belonged to no country and yet to all countries. We have gone into the ocean commons—on the waters and under the waters—carrying our technology of destruction. We have placed our weapons deep under the earth's crust. And we have borne them on airplanes around the planet, claiming the air space above us as our national possession, our national sphere. Never mind that as the earth turns and the winds move, the air breathed in Moscow and in Washington, D.C. is the same air; the outer space in which both cities and all cities on the planet subsist is the same space. And now some are proposing to take our weapons and our national egos into space and claim it too as our national possession.

In taking possession of the earth, and now of outer space, in claiming it as *our* possession, who has thought of God? Who has thought of the community of peoples? To whom does the earth belong? To whom the oceans? To whom outer space? Who speaks for God, the creator of this magnificence? Who speaks for the children of tomorrow?

Changing Our Minds

As Einstein noted, the atom bomb changed everything but the way we think. The most important step to national and world security now is a fundamental change of mind.

The two superpowers—with their aligned states in NATO and the Warsaw Pact—have each been basing their security on an unworkable premise: that the only way to survive is to be willing and ready to destroy the other, and to take the gamble of miscalculation that could destroy the other. *In fact, the only way for either of them to survive is to cooperate in saving each other.*

Lessons from Outer Space

As we journey out to the stars, as we will and must, we need to learn this most fundamental lesson: The universe is not a system of divisible and separately acting atoms, but rather of interdependent, dynamically interacting parts. Space is not something that divides planets and solar systems, galaxies and constellations, but the place where the dynamic interaction, the bonding and communication between parts, goes on. Space is not a place where planets, solar systems, and galaxies function individually and in isolation, but where they move in eternal relationship. Space is not emptiness, but a place of energy, creativity, and communion. It is not a place that can be possessed or owned or controlled. It is a place of discovery and becoming where the old separations and divisions are revealed as illusion.

We are approaching the gateway to outer space, fearful, hopeful, and uncertain, having to choose whether we will take with us the weapons for war or the tools for peace. At such a time we might ask the question Luke asked Yoda at the beginning of this journey into inner space: *What is there?* And if we listen, we may hear the faint whispers of Yoda answering, "*Only what you take there.*"

If, on our sojourn into outer space, we are to learn the lessons it has for security on earth, we need to leave behind our weapons, our old worldviews and our old habit, war. Most importantly, we need to see space not as a new frontier for warfare, but as a new frontier for a sacred journey—a journey into community and into our own deepest essence. We must not allow that journey to be desacralized.

Repeat: What we do in outer space will be a mirror image of our inner space. It will reflect the state of our souls.

1. As quoted in "Space Militarization Becoming Space Weaponization," an issue paper by the Institute for Space and Security Studies (ISSS) (Potomac, Md., 1984), p. 1.
2. Ibid.
3. Ibid., pp. 1–2.
4. Ibid., p. 2.
5. Officially titled the *Treaty on Principles Governing the Activities of States in the Exploration and Use of Outer Space, Including the Moon and Other Celestial Bodies*, but commonly referred to as the Outer Space Treaty.
6. "The Trouble with Star Wars," an ISSS issue paper, p. 1.
7. President Reagan in his televised address to the nation, March 23, 1983.

8. Daniel O. Graham in a letter to Archbishop William D. Borders dated November 2, 1982, in which he proposed a space defense as an alternative to the policy of Mutual Assured Destruction.
9. Greg Fossedal, "Exploring the High Frontier: New Defense Option Would Stifle Soviets, Help Chances for Peace," *Conservative Digest* (June 1982).
10. Heritage Foundation, "BMD & Arms Control," as quoted by Robert M. Bowman in "Star Wars and Arms Control," an ISSS issue paper (January 1985), p. 3.
11. Fossedal, op. cit.
12. Ibid.
13. Edward Teller, as quoted by Charles Mohr in "Reagan Is Urged to Increase Research on Exotic Defenses Against Missiles," *New York Times* (Nov. 4, 1983).
14. As quoted by *Defense Monitor*, Vol. XIII, Number 5 (1983).
15. Ibid.
16. Ibid.
17. "Cost of Missile Defense Put at $70 Billion by 1993," *New York Times* (Feb. 10, 1985).
18. In 1982 the Pentagon estimated that a "damage denial" system could cost about $500 billion. See Union of Concerned Scientists, "The New Arms Race: Star War Weapons," Briefing Paper No. 5 (Oct. 1983).
19. Schlesinger's estimate was reported by William J. Broad, "Star Wars Research Forges Ahead," *New York Times* (Feb. 5, 1985). See also: *Defense Monitor*, Vol. XII, Number 5 (1983); and Robert M. Bowman, "The Reality of Star Wars," an issue paper of the ISSS.
20. "Group of Top Scientists Close to Government Fighting Space Weapons Plan," *New York Times* (Nov. 16, 1983).
21. "The Reality of Star Wars," p. 5.
22. Bill Keller, "Air Force Seeking More Wily Missile," *New York Times* (Feb. 11, 1985).
23. "The Trouble with Star Wars," p. 8.
24. Ibid., p. 9.
25. Ibid.
26. See Carol Rosin, "How the U.S. Administration Is Preparing for Star Wars," *Breakthrough* (Winter 1985), a publication of Global Education Associates, East Orange, N.J., p. 5.
27. UCS, Briefing Paper No. 5.
28. Robert Bowman, "The Case Against Space Weapons," an ISSS issue paper (January 1983).
29. Ibid.
30. Ibid.
31. Ibid., p. 5.
32. Ibid.
33. Hannah Arendt, *The Human Condition* (Chicago: The University of Chicago Press, 1958).
34. Ibid., p. 1.
35. See Arendt's excellent chapter on "The Reversal of Contemplation and Action," ibid., ch. 41, pp. 289–294.
36. As quoted by Carol Rosin in "How the U.S. Administration Is Preparing for Star Wars."
37. Op. cit.

To the extent that Bundy and his colleagues and Mische are correct, putting weapons in space would be disastrous. Is there any way that space can serve our needs? At present, the earth is watched by many satellites, most of which belong to the U.S. or the Soviet Union. They serve to tell each superpower what major weapons the other is testing or deploying, and to provide a warning of any attack. Since missile silos are increasingly vulnerable to attack, a capability to spot an attack is reassuring.

Thinking beyond deterrence, Howard Kurtz proposes using space satellites to inform not only the superpowers, but *all* nations and, in particular, a "global safety authority" that would gradually come to protect everyone. A former airline executive, Kurtz watched as air traffic control systems were designed and developed. Without them, aviation would lapse into anarchy, but with the help of these systems, flying is safer than driving. Kurtz wondered how we could do the same for conflict—gather timely data, establish rules, punish violations.

One way in which Kurtz has expressed his ideas is by drafting a speech that he first hoped a U.S. President might deliver. The text has appeared in *CoEvolution Quarterly*, which was one of our favorite periodicals and which became the *Whole Earth Review*. Now Kurtz is offering the speech more broadly, to any world statesman. This may sound grandiose, but one of his ideas, for an international satellite monitoring agency, has already been proposed to the U.N. by Giscard d'Estaing, then President of France.

Speech in Search of a Statesman Capable of Delivering It

Howard G. Kurtz

I AM ADDRESSING MANKIND TONIGHT AND WILL USE every means at my command to have this message brought home to every man, woman and child in every country on Earth. What I have to say may be a matter of life or death for every human being. I hope they listen carefully. This is a message of hope.

In 1945 when the United Nations was chartered in San Francisco, many people believed that war had ended and mankind had entered a new era of peace. This was a tragic delusion. History's endless parade of war after war has continued. The potential horror of war has grown worse, not better. We are living in a new historic age of anxiety.

In 1945, that same year, the first atomic bomb was exploded. This signalled the dawn of an era of exploding research and development and escalating progress in science and technology, in hundreds of new fields. Civilization has been projected into the Nuclear Age, and then into the Space Age, and now into even newer technological ages.

As a result, the strategic power under the command of the head of state of a modern major nation is beyond easy vision or description. There hardly are words to explain. But I can tell you this: it used to be thought that only God could command the power to destroy mankind in the twinkling of an eye. But today the leaders of several nations command the power to launch the war that could annihilate world civilizations.

The great rivers of new knowledge coming from the research and development centers cannot be dammed up within the boundaries of any one nation. Within ten years perhaps thirty nations will command this power of death for mankind. By that time the weapons of total annihilation will be so cheap that criminal elements may obtain this power to hold mankind for ransom. Breakthroughs in chemical and biological and nuclear warfare will magnify and compound the forces of destruction.

This is a new historic time of dread. In this new age no nation ever again will find positive national security and political independence by use of its power to destroy human life. In this new age an attempt to destroy an enemy can destroy mankind. All nations have moved into a common chamber of danger. The world is like a room filled with gas, in which men who strike matches endanger the lives of everyone in the room.

We must create new history. It is no longer safe merely to relive old history.

There will always be war and aggression between nations until there is created a global safety authority of some kind, with the strength and the authority to maintain the common good of national security and political independence for all nations.

Historians and political scientists find no precedent for such a Global Safety Authority. But before 1940 there was no precedent for atomic bombs, as before recent times there was no precedent for many of the miracles of modern medical science.

We live in an age that breaks precedents. We need not be enslaved by historical precedents of war and catastrophe.

It is predicted that by the year 2000 world population will exceed six billion people. Scientists and engineers tell us that we have within reach the means to produce the food, clothing, housing, warmth, health, education and welfare for all of these people. But nations are not producing to meet these human needs, because today the world spends nearly $200 billion a year on an escalating arms race, producing anti-human forces of destruction.

Nations have learned the secrets of the power of death for mankind because they have invested their human and natural resources in massive mobilizations for the specific purpose of learning these secrets.

I say to every man, woman and child in every nation that the time has now come for a massive mobilization of human and natural resources for the purpose of learning the secrets of the power of safety and well-being for the people of all nations. If we believe that the intention of creation is toward greater life for the family of mankind, then let us begin to direct the forward march of science and technology toward such a greater purpose. I believe in the future of mankind.

I hold before the world the long-range vision of a stronger United Nations. At the base of this future United Nations will be a Global Safety Authority. The purpose of this Authority will be to maintain positive control over all power of destruction. It will have power to maintain public inventory of all war material and personnel. It will have authority to act to stop unauthorized war production. It will have power to maintain public inventory of all movement toward mobilization for war. It will have authority to stop such mobilization. It will have power and authority to prevent violation of one nation's borders by the forces of any other nation, large or small. It will have authority to divert any inter-nation conflict to proper courts of inter-nation conflict law.

I hold before the world the long-range vision of a stronger United Nations in which man will maintain dominion over the forces of destruction. Instead of living in ever-increasing fear of the anti-human force of advancing science and technology, man's increasing knowledge will be redirected to feed the hungry, clothe the naked, heal the sick, educate the illiterate, and free the patriotic people of every nation to work toward their own national goal in their own national way, free from the dread of war or domination by any foreign power.

Will it ever be possible to trust the national security and political independence of the United States to a stronger United Nations with monopoly control over global armed forces management? Will it ever be possible for all nations to trust their national security and political independence to such a remodeled United Nations? Will the American people ever find reason to believe that this global United Nations power will not come under the control of communists, or any other group with intentions to dominate the world? Will the people of all other countries ever find reason to believe that this global United Nations power will not come under the control of American political power?

I do not know the answer to these questions. I am proposing that a long period of serious research and exploration begin today, in search for the answers to these questions about the future safety and well-being of mankind.

To the rising generation of young men and women in every nation we declare open for exploration and pioneering a new generation of creativity and invention to dwarf any era in past history.

I say, let the national security leadership of all nations begin thinking and experimenting beyond the strategy of defense, and beyond the strategy of deterrence, to explore a new future strategy of prevention more appropriate to the age of global danger. Let us create new history more in keeping with the danger man has created.

Black and terrifying, the escalating danger of global war today looms above mankind like a thunderstorm of death. But beyond that dark horror, I see a white light of hope. I believe that man can create global safety systems in time.

I propose four giant strides forward on the road to the permanent end of war between nations . . . to lasting peace and prosperity. I propose four giant strides forward in the creation of a demonstration model of a Global Safety Authority. These will be four giant strides forward toward a stronger, more effective, and greater United Nations of the future. One day, when and if our efforts succeed, the present United Nations will evolve into a world security organization capable of guarding the national security and political independence of every nation.

I am preparing to issue directives to the appropriate departments of my government, and preparing to seek the full support of the Congress and of the American people, for four large-scale development projects:

First Stride: All-Nation Space Stations

I propose construction of a new series of reconnaissance satellites. The use of these surveillance space stations will be made available to all nations. All nations will be invited to develop surveillance systems for experimental installation in these orbiting laboratories. All nations will be given full access to all information inputs and outputs from these global surveillance systems, so that all national defense establishments, all regional defense establishments, and the United Nations Military Staff Committee can gain protracted experience in working together to test the effectiveness and to establish the future requirements for a Global Safety Authority intelligence capability. Through these satellites we may begin to find answers to the question: Can a future United Nations Safety Force be made dependable and effective?

Second Stride: All-Nation War Alarm System

I propose the construction of an advanced global command and control headquarters, near the United Nations Headquarters, with duplicate synchronous display centers in the capitals of all cooperating nations and in the headquarters of all regional defense organizations. This center will be open to the public of all nations and to the governments of all nations. The large illuminated display walls in this prototype global command center will be connected by direct communications lines to the national command center in the Pentagon in Washington. The people of the world can begin to see the global surveillance and intelligence capabilities that the United States already possesses and that enable it to exert command and control over military forces spread across more than half the earth. No information which would be detrimental to the security of the United States will be passed to these public exhibitions, but the world can begin to see the emergence of global command and control systems that could eventually bring armed forces management for the world to the new United Nations Global Safety Authority. Every other nation will be invited to connect its national command center to this world display. No other nation will be expected to divulge information that could jeopardize its own security. But all nations will gain practical experience in the interface problems in the successful operation of a future war safety control system of global dimensions. The people of all nations will begin to see the open experiments which may one day lead to the development of a new system of security.

Third Stride: War Prevention Conferences

I propose a continuing series of multinational War Prevention Conferences. Military and political leaders of all participating nations can begin to hammer out the requirements and force structures for the future Global Safety Authority. They can analyze and experiment with unprecedented new political checks and balances which will be required to make certain that this all-nation safety constabulary cannot become a tyrant over the minds of men or over the governments of independent nations, and that it cannot be captured by any one political power clique to be used to conquer the world. These War Prevention Conferences will be held on many levels, outside of official channels of the United Nations. They will be held with individual nations, or with groups of individual nations, or with regional defense organizations. Free discussion will be encouraged from every source where new ideas may emerge to help solve this problem which has been insoluble throughout the ages. Can a Global Safety Authority be made effective and safe to the point that the people of all nations will trust their national safety to this mobilization for the prevention of war? The world will be waiting for answers to this question.

Fourth Stride: All-Nation War-Prevention Games

I propose a continuing series of multinational war-prevention games. Where war games are used to develop military proficiency in waging war, these war-prevention games will be utilized to develop and test the increasing proficiency in preventing war throughout the world in the new historic era. A purpose of these games will be to allow the public and press of all nations to see that nations are making serious efforts to solve the problem of war, and what progress is being made toward this great and grave objective.

I repeat that all nations are invited to cooperate in these four giant strides forward toward peace and prosperity for all nations.

But the United States offers to lead in this ten-year demonstration no matter what nation or nations hold back at this point. We expect nations to be suspicious of our motives. We invite nations to work with us to satisfy these suspicions. We will earn their confidence.

We will use every means at our command to see that the people of all nations are kept fully informed as to the progress being made toward the global protective systems which one day will bring peace and prosperity to all nations.

No nation will be expected to weaken its present defense force or posture during these experiments. No patriotic people will be asked to give up their pride in their own native land, or swear allegiance to a massive single world government. National governments once again will concentrate on the prosperity of their citizens, instead of on preparations to kill and be killed in war with their neighbors. There will be a common defense force for all.

I believe that every people on earth can find security and prosperity in a revised and strengthened United Nations, in time. I believe man now has the inventiveness and creativity to begin to transform the United Nations into an effective world security organization, bringing safe prosperity within reach of all nations.

Each nation can work out its own independent path toward its own independent vision of a great society. This can become a great planet. I hope that every man, woman and child around the world who hears this message will tell his neighbors and his friends that the time now has come to take giant strides forward toward peace and prosperity for all nations.

Cooperating in Outer Space

In 1945, Arthur C. Clarke had the idea for communicating through a space satellite in geosynchronous orbit. Author of almost fifty books, he is most celebrated for *2001: A Space Odyssey,* for the movie of which he shared an Oscar nomination with Stanley Kubrick in 1968. A resident of Sri Lanka, Clarke addressed the U.N. Committee on Disarmament in Geneva in 1982. In the following excerpt, he endorses the idea of an international satellite monitoring agency, which would break the existing but rapidly eroding monopoly of fine-resolution space photographs held by the U.S. and the Soviet Union. To many the idea sounds crazy, as Clarke's proposal for satellite communication undoubtedly did at the end of World War II, when bomber pilots were still depending on propellers.

Everybody's Satellites

Arthur C. Clarke

IT IS NO LONGER TRUE THAT WARS BEGIN IN THE MINDS of men; they can now start in the circuits of computers. Yet the technologies which could destroy us can also be used for our salvation. From their very nature, space systems are uniquely adapted to provide global facilities, equally beneficial to all nations.

As you are well aware, in 1978 the French Government proposed the establishment of an International Satellite Monitoring Agency to help enforce peace treaties and to monitor military activities. This has been the subject of a detailed study by a United Nations Committee conducted by Dr. Hubert G. Bortzmeyer. The conclusion is that such a system could well play a major role in the preservation of peace.

The operational and political difficulties are obviously very great, yet they are trivial when compared with the possible advantages. The expense—one or two billion dollars—is also hardly a valid objection. It has been estimated that its reconnaissance satellites save the United States the best part of a trillion dollars. A global system might be an even better investment; and who can set a cash value on the price of peace?

However, the United States and the Soviet Union, anxious to preserve their joint monopoly of reconnaissance satellites, are strongly opposed to such a scheme. The British government is also lukewarm, to say the least.

Nevertheless, we have seen that in matters of great, though lesser importance, such as international communications, it is possible to have extremely effective co-operation between a hundred or more countries, even with violently opposing ideologies. Intelsat is a prime example, as on a smaller scale is Intersputnik; and in the near future Arabsat will establish its regional space system.

I like the name Peacesat, and although that has already been pre-empted by the Pacific Radio Network, I will use the term, with due acknowledgement for the remainder of this talk.

Reactions at Unispace 82 and elsewhere suggest that the Peacesat is an idea whose time has come. Those who are sceptical about its practicability should realize that most of its elements are present, at least in rudimentary form, in existing or planned systems. The United States has made its Landsat photographs, which have a ground resolution of roughly 80 meters, available to all nations. Not surprisingly, there has been some concern about the military information that these photographs inevitably contain. That concern will be increased now that Landsat D has started operations with a resolution of 30 meters; I was stunned by the beauty and definition of the first photographs when they were shown to us at Unispace a few weeks ago. The French SPOT satellite will have even better resolution (10–20 meters) and this is rapidly approaching the area of military importance, although it is nowhere near (perhaps by a factor of one hundred) the definition of the best reconnaissance satellites under favorable conditions. Whether the superpowers wish it or not, the facilities of an embryo Peacesat system will be available to all countries in the near future.

May I remind my Russian and American friends that it is wise to cooperate with the inevitable; and wiser still to exploit the inevitable.

Peacesat could develop in a non-controversial manner out of what Howard Kurtz, their long-time advocate, has called the Global Information Co-operative.

This could be a consortium of agencies for weather, mapping, search and rescue, resources and pollution monitoring, disaster watch, information retrieval and, of course, communications. No one denies the need for these facilities. If they were provided globally, they would inevitably do much of the work of a Peacesat system. The only extra element required would be the evaluation and intelligence teams needed to analyse the information obtained.

The organization, financing and operation of a Peacesat system has been discussed in the UN report, to which I refer you for details. It is not a magic solution to all the problems of peace; there is no such thing. But at least it is worthy of serious consideration, as one way of escape from our present predicament.

What would it do to the world if everybody could see everything that can be seen from above, at a resolution of thirty feet, a yard, or even a few inches? The first of these distances is close to the resolving power of a new French civilian satellite called SPOT. As far as we know, a few inches is roughly the power of the best U.S. intelligence satellites. Maybe before long a civilian or international satellite will be able to spot objects only a yard across. What will be the effects if this data is made available, in a timely way, to all governments or even to any buyer?

The Age of Transparency

Kevin Sanders

THE POTENTIAL PEACEKEEPING ROLE OF INTERNATIONAL earth-observation satellites was first reported in "Draft of a Proposed Speech for a President of the United States" by Howard and Harriet Kurtz in *CoEvolution Quarterly* #20, Winter 1978. With the advent of the first civilian high-resolution satellite, such a monitoring system has suddenly become feasible. But there is also talk in the Pentagon of shooting it down.

An Ariane rocket recently blasted off from French Guiana in South America, launching into orbit a new European satellite called SPOT (Systeme Probatoire d'Observation de la Terre). It is the world's most powerful and versatile civilian earth-observation system, producing pictures of the earth's surface three times more detailed than those of the U.S. Landsat. Under the 1984 Space Commercialization Act, Landsat has recently been transferred from the U.S. government to RCA and Hughes Aircraft, who will operate it commercially under the name EOSAT (Earth Observation Satellite). But among some U.S. officials there is concern that SPOT has already rendered Landsat obsolete. Reports from Europe suggest that SPOT may also mark the beginning of what Daniel Duedney of the Worldwatch Institute has called "The Age of Transparency," in which all nations can see everything all the time. Such a development cuts at an oblique angle across traditional assumptions of nation-state sovereignty and carries important long-term implications for the superpowers' defense and foreign policies.

The Pentagon currently imposes restrictions on Landsat pictures, limiting them to a resolution of 30 meters (one hundred feet). That means Landsat is not allowed to register anything smaller than 100 feet wide. (As a result, the public has access to considerably more detailed satellite pictures of the surface of the moon than of the earth.) SPOT, a joint enterprise by French government and commercial interests—in association with Belgium and Sweden—is not subject to Pentagon restrictions. It has been designed to detect anything more than 33 feet wide. While Landsat pictures cover a larger area—180 miles square—SPOT will reveal more detail in its pictures, covering an area 36 miles square. Landsat can distinguish blocks of houses; SPOT will distinguish the individual houses.

Jim Kukowski, head of NASA's Space Science Information office in Washington, D.C., says, "The impact of SPOT on Landsat's commercial viability will be significant. Landsat cannot be changed; it was built to certain specifications. We in the civilian sector can use our sensing devices to get down to a certain resolution where we have to stop. At that point it comes under the umbrella of the Department of Defense." He also notes that some of the unclassified pictures from the U.S. space shuttle orbiter's photographic cameras already exceed Pentagon restrictions on Landsat. "In some orbiter pictures we can pick up a 747 in flight," he says. As a result of the ten-meter resolution available from SPOT, another NASA official warns, "We have to ask whether we are giving the private sector a dead duck with Landsat."

Congressman George Brown (D-California), who has long advocated civilian access to high-resolution pictures of Earth from space, predicts, "The U.S. probably will not be able to compete with SPOT, either in the quality of its pictures or the international marketing of the service." But Gilbert Weill, President of SPOT Image (the corporation that owns and operates the satellite), stresses that SPOT pictures can be used in cooperation with Landsat. "SPOT is a logical complement to Landsat," he says. "We offer detailed images that can work in tandem with their more broad-scale product."

At the National Oceanic and Atmospheric Administration (NOAA) in Washington, D.C., an official who asked not to be identified, acknowledged that U.S. reaction to SPOT is schizophrenic. "Even though SPOT will be competition for Landsat, cooperation makes sense; the two satellites gather somewhat different kinds of data serving different needs. Also, we don't know if the present Landsat will continue working until we get the next one in orbit." The next in the series, Landsat 7, is scheduled for launch by the U.S. government in 1988 and will be operated on a commercial basis by a consortium of private space-technology companies. An NOAA official reported that future Landsats have been cleared by the Pentagon to go down to a resolution of 15 meters. SPOT will still be well below state-of-the-art for earth observation from space. Already some of the approximately 100 U.S. and Soviet spy satellites now in orbit are believed to have a resolution of one meter or even less. Anecdotal reports claim that some can see any weapon larger than a rifle, and that on a clear day they can detect the magazine you are holding. Soldiers working on secret projects in the field are ordered not to shine their shoes; the glare from polished shoes marching in step is said to show up in military satellite pictures. General Daniel Graham (ret.), who heads the Washington, D.C.-based Heritage Foundation's High Frontier proposal for a space-based weapons system, boasts, "From space we can tell on which side a man's hair is parted."

Over the years the superpowers have learned to coexist under each other's scrutiny from space. Both concede that spy satellites have helped stabilize relations and reduce tensions. The advent of SPOT, however, undermines the superpowers' monopoly on high-resolution pictures from space, and there are now plans under consideration in Europe to experiment with SPOT data to monitor crisis areas, military activities, and arms control agreements.

International satellite monitoring of peace agreements was first proposed in the early sixties by a Washington D.C.-based group called War Control Planners, headed by airline executive Howard Kurtz and his late wife, Harriet, a theologian. Edward Teller, a developer of the hydrogen bomb, was an early supporter. "Everything that can be seen from space should be shown in the United Nations," Teller said. The idea finally emerged in 1978 as a French U.N. proposal for an International Satellite Monitoring Agency (ISMA), after being picked up by assistant secretary general of the U.N., Robert Muller.

After a 14-year, 12-nation study, the U.N. issued a report in 1982 concluding that an ISMA was "feasible and desirable." According to the U.N. report, each year an ISMA would cost the international community "well under one percent of the total annual expenditure on armaments" and could be operated "with or without the support of the superpowers."

Although the U.S.S.R. voted against the ISMA, and the U.S. abstained, 126 nations voted in support. In 1983, the Parliamentary Assembly of the Council of Europe—a 23-nation body including the Vatican—offered to cooperate with the U.N. through the European Space Agency (ESA), to establish a rudimentary ISMA. They would use data routinely available from U.S. and U.S.S.R. civilian satellites, together with more detailed pictures available from SPOT. The project is currently under further study by a Parliamentary Assembly committee on science and technology. According to John Pike, a space-science researcher at Federation of American Scientists in Washington, D.C., the Europeans are eager to get involved with an ISMA to reduce their dependence on earth observation material they currently get from the U.S. military in exchange for other "intelligence" information. "A lot of European governments would like to get out of that deal," Pike says. Howard Kurtz, who sees an ISMA as the first step to what he calls a "global information cooperative," regards the advent of SPOT—and its possible value to ISMA—as most encouraging. Kurtz is also urging support for a congressional resolution to guarantee limited civilian access to NAVSTAR, the Pentagon's multibillion-dollar, 20-satellite, global navigation and tracking system due to be in place in 1988. "As part of an ISMA, it (NAVSTAR) could be used for global and local security, earth resource management and crisis and disaster relief," Kurtz says. "Back when Harriet and I first proposed these ideas, everyone said we were 25 years ahead of our time," he recalls. "Well, it's been 25 years and the time seems right." (Kurtz has been nominated for the Nobel Peace Prize for his work.)

In a 1982 address to the U.N., British science writer, Arthur C. Clarke—who has dubbed the ISMA "the peacesat"—said it was an idea whose time has come. "Most of its elements are already present in existing or planned systems. The French SPOT with a ten-meter resolution has been mentioned. Whether the superpowers wish it or not, the facilities for an embryo peacesat will soon be available. May I remind my Russian and American friends that it is wise to cooperate with the inevitable."

Despite the current controversy over the Reagan Star Wars proposal—Canada, the Netherlands, France, Norway, Denmark, Sweden, China, Australia and New Zealand refuse to cooperate in research—the degree of international peaceful cooperation in space is, by contrast, already considerable and growing. Given the inherently global nature of space, many believe such cooperation is inevitable. All nations already cooperate in the space-based telecommunications system, Intelsat. The European Spacelab—involving all 12 nations of ESA and Canada—flies on the U.S. shuttle. Australian tracking facilities will follow the European Halley's probe. France, Canada, the U.S. and the U.S.S.R. are involved in the Search and Rescue Satellite system (SARSAT). Like the international postal service and civilian air traffic controls, satellite-based systems for communication, meteorology, astronomy and earth observation require a high degree of international cooperation to function efficiently. For example, the SPOT scanning programs are directed from mission control in Toulouse, France, yet the digital data from which the pictures are reconstructed will be collected through a network of receiving stations in 12 countries, with more soon to be involved. Already 48 nations have ordered test pictures.

At a meeting in Washington, D.C., last year, Dr. Caesar Voute, chief of the Netherlands-based International Institute for Aerial Survey and Earth Sciences, predicted that, as such cooperation expands, "Interdependence will take the place of detente and coexistence." Dr. Voute regards the ISMA as an opportunity for all nations to work together for common global security. "We are at an historic turning point," he said, "but it will require a new approach by mankind to reap the benefits of space."

Cooperating in Outer Space

Ironically, President Reagan's proposal to share Star Wars technology with the world would entail the greatest international cooperative endeavor ever undertaken. Armand Hammer, president of Occidental Petroleum and friend of both President Reagan and Secretary Gorbachev, has challenged President Reagan to offer an immediate exchange of space weapons technology. But historian William Irwin Thompson has mocked what he regards as the inherent absurdity of the idea, arguing that, paradoxically, it would also require such a massive level of trust in the exchange of nuclear, space and computer technologies "with Rockwell subcontracting space weapons construction to the Russians" that a situation would soon be reached in which it would be easier, cheaper and safer simply to disarm the bombs on earth and secure space for peaceful purposes.

But even purely civilian spacecraft present ambiguities of conflict and cooperation. Considerable international confusion exists over the military implications of satellites. Earlier last year, just before he left for Geneva to begin the first round of arms negotiations with the U.S.S.R., U.S. chief negotiator Max Kampelman astonished a group of space-cooperation activists by telling them, "There are already thousands of weapons in space." After some enquiries he modified the number to "hundreds." A few days later, after being challenged further on the statement, Kampelman explained that he was referring to the military observation and communication satellites. A similar confusion seems possible over SPOT, since its data could also be used to identify military movements and installations. Would this make it a "weapon"? Are there circumstances in which the U.S. would consider using antisatellite weapons against SPOT or other components of an ISMA?

The Pentagon is reluctant to discuss the issues. "We don't want to reveal the range of our ASATs (antisatellite weapons)," said U.S. Air Force information officer Ron Rand. But Jonathan Weiner, aide to Senator John Kerry (D-Massachusetts), who opposes ASAT testing, says, "SPOT is clearly vulnerable, particularly if nations develop laser weapons." A Pentagon official acknowledged that if SPOT were to reveal military activities the U.S. wanted to keep secret, and if France refused to withhold the pictures, "It would be a perfectly logical scenario that SPOT could be targeted." David Julian, SPOT Image vice president, says the billion-dollar SPOT system carries no defense, and adds, "The possibility of an ASAT attack has not been discussed." He speculates it would be, "unlikely, since existing military satellites can already see much more." But Congressman George Brown claims, "The development of ASATs will inevitably pose a mortal threat to the whole international civilian space enterprise." (Currently Congress has ordered a halt to U.S. ASAT weapons testing unless the Soviets resume their program.)

Possibly in anticipation of such tensions, French President Mitterand has proposed an independent European space station be used to "observe, transmit and counter any eventual menace." He suggested recently that the French-built civilian science and technology platform, Eureca, which is due to be launched in two years, could be adapted for European space defense. Some observers believe the French statements merely reflect European resentment of President Reagan's failure to consult other nations prior to his public announcement of the Star Wars project.

However that may be, other nations are developing an increasing stake in space, and a growing concern for its security. Four more SPOT satellites will be launched in the next ten years, and West Germany, Japan and ESA also have earth observation satellites in the works for later this decade—some with picture quality and resolution even better than SPOT's. Charles Sheffield, vice-president of the privately owned, Washington D.C.-based Earth Satellite Corporation, predicts the trend will continue. "By the year 2000 it is hard to imagine that there will be any limits on resolution of spaceborne sensors, other than those imposed by the technical state-of-the-art of the future optical systems," he says.

Eventually a network of such satellites—perhaps in the form of an ISMA—may become what science writer Ben Bova calls "a Swiss guard in space." Some believe such a system could pose an international counterforce to the growing threat of the weaponization of space. It could herald an age of international governance with a third force in orbit to triangulate the political dynamics of space with the superpowers.

Cooperating in Outer Space

The coming age of transparency is likely to require a new international etiquette and will raise a number of questions that have, as yet, been little considered: Will governments respond with greater civility or greater stealth? Will national and even personal privacy yield to the requirements of world security or have they already? Will orbital space become a superpower battlefield or a global commons? These are but a few of the issues the superpowers may have to consider now that the giant eyes of SPOT are in orbit, sending back from space the best pictures we have ever been allowed to see of our planet. They are available from SPOT without restriction for about a dollar an acre. Extra for 3-D.

In the following paper, originally published in *Foreign Policy,* Daniel Deudney summarizes and extends many of the themes in this section. In his critique of Star Wars, he adds to the arguments advanced by Bundy and his colleagues and by Mische. In his proposals for peaceful and cooperative uses of space, he works in the spirit of Kurtz, Clarke, and Sanders. Written when Deudney was a senior researcher at the Worldwatch Institute, this paper clearly samples both the dangers and benefits possible in space. As it now stands, we're in a race to militarize space, as we have militarized almost every other medium. Here are the consequences that would follow if we continue, and here are some of the alternatives available to us. It's our choice, if we can induce the Soviets to agree.

Unlocking Space

Daniel Deudney

"GIVE ME THE POINT OF MY CHOOSING AND I WILL leverage the world," said Archimedes. Outer space is both the leverage point for the current superpower military competition and the point from which an alternative security order could be built. In fact, despite the futuristic aura that surrounds space and space technology, the strategic military balance has, at least since the advent of ballistic missiles in the late 1950s, centered on space technology.

Today the world is at the edge of perhaps the most far-reaching military threshold since the beginning of the atomic age. Today, as then, a technological breakthrough both threatens unprecedented destruction and provides the opportunity to create a fundamentally more secure international order. Space weapons have already destabilized the nuclear arms race and may fatally upset the strategic balance. Yet a ban on space weapons would literally put a ceiling on the arms race and block a costly and destabilizing new arms competition. Perhaps more important, it would preserve the incomparable vantage points of space for the monitoring platforms and joint scientific enterprises of a planetary security system.

Tragically, these opportunities to head off an extraordinarily expensive arms race and construct an alternative security system are about to be lost. Military utopians preach, despite all historical evidence, that true security resides in the next technological advance. The temptation to exploit emerging technologies for ephemeral unilateral military advantage is strong. History and logic suggest, however, that technical innovations will broaden and intensify, not end, the arms race.

Although civilian space programs capture the lion's share of public attention and until the late 1970s the U.S. space budget, the superpowers' military activities have always dominated the use of space. As of 1983 the Soviet Union had successfully launched 1,017 military and 521 civilian missions, while the United States had sent up 440 military and 356 civilian payloads. The Soviet lead in military missions reflects the longer lives of U.S. satellites, not a menacing space gap. These numbers are only approximations because some flights are dual-purpose and because the Soviet Union reveals little about the intent of its launches.

As the popularity of prestige space spectaculars has fallen, along with support for Soviet-American détente, the military has won an increasingly large share of space outlays, finally surpassing the civilian budget in the late 1970s. Taking into account military activities hidden in such dual-use programs as the space shuttle, the military today accounts for close to 75 per cent of U.S. space spending. Similar budgetary priorities characterize Soviet space spending. The U.S. Department of Defense estimates that military programs make up 70 per cent of Soviet space launches, while a further 15 per cent of launches combines civilian and military programs.

Moreover, the line between civilian and military space technology is exceedingly fine, if not altogether artificial. The civilian space programs of the 1960s evolved from the military advances of the 1950s. A military missile differs from a civilian rocket in payload and target, not basic technology. When asked to explain the difference between the Atlas rocket that sent astronaut John Glenn into orbit and those poised to obliterate the Soviet Union, President John Kennedy reportedly replied, "attitude." Communications, remote sensing, and navigation and weather satellites pioneered by the military also possess this dual character. Because civilian and military space technology are practically indistinguishable, no regulation of military capability can succeed without a carefully designed regime for civilian activities.

Understanding how space technology has and could shape the strategic balance requires reviewing the geography of space. Earth's atmosphere trails off into almost nothing within 100 miles of its surface, which means that most people live closer to space than to their national capitals. National sovereignty, however, cannot reasonably extend from airspace into near-orbital space, for nothing can achieve and maintain orbit above just one country. In this respect, space is like a ball of string—the orbital threads can be wrapped around in any pattern, but if the webbed ball is cut in half, its value is lost. And while the long-range political consequences of space's geography will become clearer as technology advances, one conclusion appears likely: one state's effective control of space and its ability to prevent the passage of another state's vehicles through space could bring planet-wide hegemony. Thus far, the logistical difficulties and expense of launching and maintaining space vehicles have obscured this potential. But as long as the military potential of space technology remains unrestrained, the political fate of space will cloud the future of independent societies on Earth.

The High Ground

Military exploitation of space has passed through three phases: the first, marked by missiles; the second by reconnaissance and surveillance satellites; and the current phase marked by space-based force multiplier satellites that enhance weapons capability. The first use of outer space as a friction-free corridor for lightning-fast, unobstructable attack remains the most important. Although the United States fielded a reliable nuclear rocket force before the Soviet Union, opening this new domain permitted the Soviets to offset U.S. domination of the air.

The correspondence for countries of nuclear weapons ownership and space-launch capability betrays the prime motive behind national space programs. Of the six countries that have built space launchers, only Japan does not also possess nuclear explosives. The last two countries to enter the space club—India and the People's Republic of China—are the last two countries to explode nuclear weapons. And it is surely no accident that Libya's well-known quest for atomic weapons was paralleled by a less well publicized effort to lure West German missile scientists to that country. While largely ignored, the proliferation of rocket technology poses a serious threat to international security. Just as the implausible vision of "atoms for peace" furthered the spread of the bomb, so too the vogue but vague "space for development" movement legitimates the spread of intercontinental ballistic missile (ICBM) technology.

Near-orbital space has also been aptly called the high ground of the planet. Like a hill on a battlefield, space is not only an easily defended location from which to launch a surprise attack, it is an ideal observation point. Not surprisingly then, the information-producing satellites are the second most important type of military space vehicle. They include surveillance, navigation, communications, damage-assessment, and early warning satellites. The United States has benefited most from these reconnaissance satellites, for Soviet society is much more closed and secretive than American society.

Since the early 1970s the United States has relied upon the boxcar-sized Big Bird and KH Keyhole II satellites that scan every spot on Earth in daylight every 24 hours. The vehicles send back a constant stream of video images and periodically parachute into the atmosphere cannisters of high-resolution film, which are plucked out of the sky by specialized aircraft. Aided by weather satellites that indicate when and where to take pictures through holes in the clouds and by long-wave infrared and microwave sensors able to penetrate clouds, observation satellites record highly detailed images from many parts of the electromagnetic spectrum.

At least in their early days, satellite reconnaissance and surveillance helped stabilize the superpower military competition, encouraging both unilateral informal restraint and bilateral treaties. Satellites permitted the United States to maintain "Open Skies" over the Soviet Union after the Soviets developed guided rockets to shoot down U-2 aerial reconnaissance planes. In the early days of the Kennedy administration, the first satellite photographs of Soviet missile facilities were instrumental in exploding the missile gap, thus partially averting an expensive U.S. crash program. Several years later, President Lyndon Johnson claimed that the $40 billion spent on the space program had saved 10 times as much by reducing arms expenditures.

Indeed, the current arms control regime is based on surveillance satellites. The Soviet Union has abandoned its early claim that observation satellites were "spies" committing "espionage" and has built systems similar to American systems. Without such national technical means of verification, it is doubtful that the superpowers would have felt secure enough to negotiate and sign agreements limiting the numbers of strategic weapons.

Space-based information services have further enhanced global stability by enabling governments to monitor crises, watch for remote nuclear weapons tests, and communicate with each other quickly. Observation satellites monitor possible threats to the regimes established by the Nuclear Non-Proliferation Treaty of 1970 and Limited Test Ban Treaty of 1963, such as the nuclear test facility constructon site spotted in South Africa by the Soviet Union in 1977. After intense diplomatic pressure, South Africa dismantled the test site, thereby averting a dangerous addition to the nuclear club. Intense reconnaissance satellite activity took place during the 1971 Indian-Pakistani war, the 1973 Arab-Israeli war, and during Iran's 1980 turmoil. The ability of U.S. and Soviet leaders to contact each other within minutes through the satellite-based hot line may someday prevent a similar conflict from becoming a superpower confrontation.

Paradoxically, satellite information systems that helped to stabilize the superpower arms race in the 1960s are now accelerating the drift toward nuclear war fighting by bolstering defense planners' faith in the prospect of accuracy in targeting and in the unjustified belief that nuclear wars can be limited and controlled.

Information satellites used to communicate, navigate, collect geodetic data, and find targets are force multipliers: they make existing weapons more deadly. These technologies have cumulatively undercut, if not altogether negated, the security accomplishments of the SALT process. Typical of the quiet yet far-reaching impact of these force multipliers is the use of satellite data to calibrate ballistic missile trajectories. The extent and consequences of these changes in missile accuracy influenced the SALT II ratification debate and were hotly debated in the window-of-vulnerability controversy. Several analysts have claimed that the vulnerabilities of the U.S. land-based missile force were exaggerated because Soviet missiles had never been test fired over the Arctic Circle, where they would pass in wartime. However, geodetic satellites have mapped gravitational anomalies in these areas and trajectories have been adjusted.

The superpowers' drift toward nuclear war-fighting postures has also been spurred by and has in turn led to new investment in satellite communications systems. Particularly important have been the military's recent heavy expenditures on satellite systems to expand command, communication, and control (C^3) capabilities in order to increase control of far-flung military forces in crises and to improve communications during extended nuclear war.

Yet as is often the case in modern strategic thinking, the acronym has become a substitute for thought. Additional options, expanded communication contacts, and real-time or live information will probably overwhelm users with their sheer quantities and complexity. Local events may needlessly precipitate global crises, for local administrators may send excessively alarming reports and thereby manipulate those at the center of the network. Compare the C^3 system with the first global communications network. The telegraphs and submarine cables of the late nineteenth century were designed to improve the management of military force in the far-flung British Empire. In a detailed account of the British Colonial Office, however, historian R.V. Kubicek concluded that better communications "intensified involvement but denied . . . its corollary, control."

Today the technology of global real-time communications replaces the quasi-automatic responses of diplomats and local commanders with the unpredictable reaction of quasi-informed, overburdened heads of state.

Nor is it likely that expanded systems will make nuclear war controllable. During such a conflict, where the fog of war would be much greater and the forces to direct far more numerous, dispersed, and deadly than in a crisis, the deviation of a single nuclear-capable unit from a controlled-escalation or cease-fire plan could spell the difference between a limited nuclear war of World War II-like dimensions and the end of civilization. It is hardly reassuring that during the crisis sparked by Cambodia's 1975 seizure of the *Mayaguez* and the American military's rescue operation—often cited in military literature as a model of satellite-based C^3 effectiveness—bombing runs against mainland Cambodian targets were being launched one-half hour after President Gerald Ford ordered an end to strikes. Clearly the vastly expanded C^3 systems make war more likely by eroding the distinction between war-fighting and deterrent forces in exchange for a largely illusory ability to control wars once they have begun.

Other force-multiplying systems that tempt pre-emptive first strikes and undercut arms control agreements are now in the research or testing phases. In a move yet unmatched by the United States, the Soviet Union has launched satellites that scan the oceans with high-powered radar pulses. Linked to air- and sea-based missiles, these satellites could give Soviet commanders the information they need to launch a surprise attack on major U.S. Navy ships. The new U.S. Navstar global-positioning network of eighteen satellites and the comparable Soviet Glonass sytem will provide submarines with such precise navigational fixes that submarine-launched ballistic missiles could knock out hardened missile silos. Installing surveillance satellites with real-time data links will turn arms control monitors into war-fighting systems that can target mobile forces and assess damage following an attack.

Despite worries about the dangers of first-strike weapons, arms control efforts have neglected force multipliers. Satellite-generated information is widely seen either as inherently beneficial, inevitable, or as a good defense bargain. Yet since force multipliers shrink uncertainties—and thus weaken deterrence—these information satellites make the already urgent need for ballistic missile disarmament even more compelling.

Some technologies cannot be restrained, some can be largely demilitarized, and some can be prohibited. Geographical information, for example, is hard to get rid of once acquired and is useful for understanding problems such as earthquakes. Restrictions on satellite maneuverability, camera resolution, and real-time data flow would be hard to achieve without compromising the ability to provide early warning and verify arms control agreements. In lieu of today's practice in which each superpower deploys separate but nearly identical navigation satellites, however, a modestly accurate satellite navigation system allowing universal access for most civilian uses could be built cooperatively by the space club members. Then neither side could attack such satellite systems without denying itself their capabilities.

ASAT Leapfrog

The growing strategic importance of satellites has spawned both Soviet and U.S. anti-satellite (ASAT) weapons. Several satellite destruction techniques being investigated by the superpowers have existed in a partially tested, undeployed state for many years. New, however, is the perception that destroying war-fighting enemy satellites would be useful in an extended nuclear war or as a surrogate for terrestrial war. The exploitation of ASAT technology could open up an expensive and volatile new dimension to the arms race.

The United States abandoned a nuclear ASAT during the 1960s, when it became clear that a nuclear blast in space would not only obliterate its target but send out a pulse of electromagnetic radiation so strong that all satellites not shielded by Earth at the time of the blast would also be destroyed. Furthermore, the Limited Test Ban Treaty of 1963, prohibiting nuclear explosions in space, made tests illegal. That the United States unilaterally dismantled its nuclear ASAT system because of the test ban, however, suggests that test bans can reduce confidence in a weapon that is already judged to be of limited usefulness.

Yet the absence of a specifically dedicated ASAT system does not prove an inability to destroy satellites. Both superpowers could easily clear space with their ICBM warheads. The ASAT debate focuses instead on whether more discriminating techniques of destruction should be tested and deployed.

Since 1968 the Soviet Union has tested a specialized non-nuclear ASAT system that hovers near the intended victim and then explodes, shattering the fragile target satellite with a shower of shrapnel. The Soviet Union has tested this satellite killer several times in conjunction with large-scale test launches of ballistic missiles from silos and submarines. This coordinated operational testing, so far unsuccessful, feeds U.S. fears that the Soviet Union is preparing a first strike. Rarely in the annals of warfare has a system of such limited capability set off such alarm. Half of the tests of these hunter-killer satellites have failed. Since they are launched one at a time from large, conspicuous launch facilities and go into an easily tracked orbit before rendezvousing with their target, they are certainly not a weapon of surprise attack. Furthermore, these devices have only been tested in orbits much lower than those of U.S. early warning and communications satellites, which hover in 22,500-mile geosynchronous orbit.

The United States has designed and will soon test a new satellite killer that will leapfrog over the Soviet capability. Under this plan, a high-flying fighter plane launches a small but fast homing missile that collides with the target. Just like the Soviet system, the direct-ascent satellite killer can reach only low-orbit targets. But most Soviet satellites are in low or elliptical orbit and thus within its range. Once the homing technology has been perfected, the air-launched missile could be mated with a more powerful ground- or sea-based rocket booster, thus expanding its range.

The U.S. system will be far more destabilizing than its Soviet counterpart. Unlike the Soviet orbital rendezvous system, which can be tracked as it is launched and as it closes on its target, the U.S. direct-ascent system could strike without warning from fighter planes located anywhere in the world. A globally coordinated U.S. strike could probably destroy most operable Soviet satellites within one day.

Beyond these ASAT systems are directed-energy weapons such as lasers and particle beams. Although lasers can designate targets and determine their ranges, they are not yet potent weapons. Laser weapon prototypes cannot penetrate fog, smoke, dust, or rain; they dissipate over relatively short distances in the lower atmosphere, making them relatively ineffective even against slow-moving targets. The first laser weapon, therefore, is likely to be an ASAT weapon. If history is any guide, the next move in the deadly space arms race may be a Soviet laser orbital station designed to cripple satellites. Given technical obstacles and the Soviet tendency to deploy much earlier in the development cycle than the United States, lasers will probably be able to do little against low-flying satellites except perhaps to reduce warning time. But Soviet secrecy and inflated propaganda claims combined with the West's eagerness to believe the worst will no doubt endow the system with formidable capabilities indeed. As with *Sputnik* the psychological impact may be traumatic, leading the United States into an ambitious crash program to put battle stations into orbit.

An extensive ASAT capability will increase the likelihood of accidental and uncontrollable nuclear war. Because the eyes and central nervous system of the high-strung nuclear strike forces are already in space, every satellite malfunction will have to be treated as the harbinger of surprise attack. Whatever the usefulness of limited shows of force, it makes little sense to start a preliminary shoot-out among the systems that must remain intact to control escalation and maintain crisis communications. The Archduke Francis Ferdinand of World War III may well be a critical U.S. or Soviet reconnaissance satellite hit by a piece of space junk during a crisis.

An extensive ASAT capability will also erode the ABM Treaty. Because the difference between ballistic trajectory interception (banned by the ABM Treaty) and orbital interception (not banned) is negligible, an extensive ASAT system would probably have some ABM capability. By using technologies outlawed by the ABM Treaty, a full-scale ASAT system would reduce ABM breakout time, the period needed to convert available technology into a full-scale ABM capability. The type of high-speed collision technology being pioneered by the United States could be retrofitted onto longer-range rockets to create ABM capacity. Similarly, a Soviet laser orbital battle station with limited ASAT capability will be seen as a Soviet ABM system in the making. These doubts and uncertainties, combined with fears that the extensive Soviet air defense system may be effective against ballistic missiles, could lead to U.S. withdrawal from the ABM Treaty.

Advocates of ABM systems* speak glowingly of the strategic revolution that will repeal the "guaranteed genocide" of mutual assured destruction (MAD) and thus usher in an age of "assured survival."

* The fashionable term "ballistic missile defense" misleadingly suggests that the ability to destroy ballistic missiles cannot also be exploited for offensive purposes. For the sake of continuity and accuracy, the new, nonmissile ASAT systems should be called antiballistic missile systems (ABMSs).

Schemes for this strategic *deus ex machina* run the gamut of technical sophistication and plausibility. Retired U.S. Air Force General Daniel Graham, former head of the Defense Intelligence Agency, says that with "off-the-shelf" technology the United States could by 1988 have a global ballistic missile defense at a cost of $15 billion. The first layers of this system would consist of 432 satellites, each armed with 40 to 45 small non-nuclear missiles capable of shooting down Soviet ICBMs in the early, most vulnerable phase of their flight trajectories. Even more speculative are presidential science adviser George Keyworth's proposal to put mirrors in space and lasers on the ground, Edward Teller's X-ray lasers, and the Fusion Foundation's beam weapon aviary.

The U.S. policy toward such technological possibilities is in flux. The Defense Department research establishment has repeatedly stated that funding is adequate, demonstration premature, and the technology intriguing but a long way from the stage where it can be turned into weapons. Senior military leaders seem even more skeptical of the technology's potential and fear that these weapons' potentially voracious appetites for funds will starve vital but less glamorous terrestrial programs. The Soviet leadership publicly characterizes the new space technology as dangerous and a violation of the ABM Treaty. Although not committing the United States to do anything more than is already being done, President Ronald Reagan's Star Wars speech of March 28, 1983, encouraged ABM proponents and has prompted a U.S. policy re-evaluation that will probably result in a call for increased research and development (R&D) expenditure.

The available evidence suggests that large-scale space weapons designed to shoot down ballistic missiles suffer from tremendous, probably insurmountable, technological, tactical, and strategic problems. These systems violate no laws of nature, but the engineering requirements are daunting. Tracking and aiming technology of unprecedented precision, larger and more-perfect mirrors than any now in use, and lasers more powerful than any ever built must be fashioned to withstand the extreme temperatures of space, the stresses of launch, and the rigors of battle. A generation of scientists, tests, and R&D budgets will pass before these questions can be answered with confidence. Dr.

George Rathjens, who worked on the ABM program, says a complete system, including various ground-support and transportation components, capable of shooting down the existing Soviet ICBM force, could cost $500 billion.

On a yearly basis, a space-based ABMS would probably cost more than the current budget for the Air Force or Navy. Aside from these obstacles, an ABMS would be too vulnerable to attack and too easily foiled to be useful. Nuclear explosions, directed energy, or simple collision can easily blind the delicate sensors that track oncoming missiles. Decoys, mirrored warhead surfaces, spinning warheads, and warheads laminated with burn-resistant surfaces could fool the systems. Or, as physicist Richard Garwin has noted, a potential adversary could keep small, numerous, and cheap "space mines" hovering near a laser orbital battle station and detonate them on command. To overcome only some of these countermeasures large areas of space must be kept clear of an adversary's spacecraft—a difficult and illegal task. Even using optimistic cost figures, a dime could counter a dollar of defense expenditure—a defense bargain no country could afford.

The most plausible strategic role of a space-based ABMS would be to backstop a pre-emptive first strike. Even an incomplete or thin ABMS may preclude a second strike it if obliterates residual enemy warheads in the wake of a first strike. Unlike the various ground-based systems intended to protect fixed missile silos, a space-based ABMS might be most useful against the small number of city-destroying retaliatory weapons that would be left after a first strike. An ABMS does hold out the hope of protection against the retaliatory strikes central to MAD but it offers little hope for deterring first strikes. As long as nuclear forces exist, MAD's abolition will be a step toward, not away from, nuclear war.

Some ABMS advocates have advanced the absurd idea that America share ABMS technology with the U.S.S.R. Unfortunately, the existence of two orbiting ABMSs may also provide an overwhelming advantage to the side that strikes first. An orbiting ABMS would be highly vulnerable to attack by another of its kind. Because of the suddenness with which one ABMS could attack and destroy another, the advantage of striking first would be tremendous.

Cybernetically Initiated Annihilation

Large-scale space weapons also pose command and control problems and would be another example of what may be termed destruction entrusted automatic devices (DEADs). Perhaps the most well-known strategy involving DEADs is launch on warning. And as became clear during the debate in the late 1960s, a land-based ABM system would require computer control. As President Dwight Eisenhower's science adviser George Kistiakowsky observed, "The decision [to launch the ABM] has to be made automatically by a computer or by a comparatively junior military officer." It is unlikely that humans could ever command and control space weapons. A space laser would have about five minutes to knock out an ICBM in its slow-moving, vulnerable boost phase. It would have to respond to an attack from an orbiting laser system in fractions of a second.

The accidental triggering of a defensive system seems to be less ominous than the accidental launch of a nuclear missile. Yet with space weapons the distinction between these events largely disappears. If both sides deployed hundreds of very large, automatically controlled battle stations in orbit, a malfunctioning satellite's likely target would be another such satellite—perhaps even one of its own companion craft. Since any first strike would begin with an attack on the other side's defensive satellite system, even the inadvertent destruction of such satellites would be indistinguishable from a first strike. A malfunctioning ABM would not destroy a potential antagonist's critical military systems. Instead of removing the tightening nuclear noose, space weapons promise to bring humanity to the edge of cybernetically initiated annihilation.

Space wars will also largely preclude the peaceful use of outer space. While the universe is infinitely vast, Earth's orbital space is finite. Space junk—small bits of debris flying about in odd orbits—is already a growing problem, and it will be exacerbated by further ASAT explosions and collisions. And the X-ray laser tests proposed by Teller would damage important civilian satellites with an electromagnetic pulse (EMP). By the late 1980s over 100 communications satellites in geosynchronous orbit, worth between $5 and $10 billion, will be in jeopardy. Space presents a choice: science, information services, and Earth monitoring—or weapons.

Fortunately, because space weapons have not been extensively tested and deployed, truly comprehensive controls would still be relatively easy to devise and verify. The present space arms control regime—the Limited Test Ban Treaty, the ABM Treaty, and the Outer Space Treaty—has worked remarkably well but contains an important loophole.

Until the early 1970s the superpowers, which had not yet developed war-fighting doctrines and satellites, had agreed informally not to deploy ASATs. Yet in the late 1970s this informal restraint disappeared as the Soviets periodically tested a co-orbital interceptor as a hedge against the U.S. space shuttle and growing Chinese space assets. In response the Carter administration pursued a dual policy of negotiations and ASAT development. But the talks, which the Carter administration perceived to be stalled, were broken off by the United States to protest the Soviet invasion of Afghanistan. The Reagan administration has shown no interest in resuming negotiations despite several high-level Soviet treaty proposals and calls for talks.

American thinking on space arms control has focused on limited agendas and specific technologies, because of verification-related fears and an unwillingness to close off promising technological avenues. The Soviets, however, have tended toward sweeping bans on all space weapons or even "demilitarizing" space. The Soviets point out that a treaty banning merely ASAT weapons would force them to abandon projects they have focused on without restraining superior U.S. weapons innovation. But as long as the ABM, Limited Test Ban, and Outer Space treaties are in force, the only remaining space weapon treaty needed is an ASAT ban.

A panel of prominent scientists assembled by the Union of Concerned Scientists recently presented to the Senate Foreign Relations Committee a draft treaty that cogently described the traditional U.S. philosophy on ASAT control. Their treaty proposes a comprehensive ban on the testing, production, or deployment of any weapons system that could damage, destroy, or interfere with any country's spacecraft. It would also outlaw any weapon stationed in outer space capable of inflicting or causing any form of damage on the Earth, in the atmosphere, or on objects placed in space. A standing consultative commission would resolve any controversies arising from the treaty. This I-know-it-when-I-see-it language may effectively prevent the emergence of unanticipated activities and technologies, for in theory it could be agreed upon quickly and would allow the standing commission to fill in the substantive gaps later.

Alternatively, militarily significant capabilities—or functionally related observable differences (FRODs), as the jargon of SALT refers to them—could be expressly prohibited, thus satisfying both the U.S. demand for clear verification bench marks and the Soviet insistence on comprehensive controls. A treaty outlawing specific systems would be more difficult to negotiate but probably easier to ratify in the U.S. Senate. Moreover, even if the shorter route is taken, the United States will have to agree both internally and with the Soviets upon specific activities a general treaty bans.

The most important items to prohibit are explosions and unmanned orbital rendezvous, collisions or close high-speed passes, and close approaches in geosynchronous orbit. These provisions would effectively ban both current Soviet and prospective U.S. ASATs and space mines. Size limitations should be placed on laser mirrors, the capacity of atomic power sources, and where definable and verifiable, electromagnetic interference, thereby outlawing directed-energy weapons while permitting small-scale civilian use of these technologies.

Limits on power sources in space as an arms control measure are little discussed but potentially very valuable, both to prevent space information systems from becoming war-fighting systems and to block directed-energy weapon use in space. Nuclear reactors alone can power advanced force multipliers because they provide much more electricity than do photovoltaic cells and are much sturdier and more maneuverable than solar-powered satellites. A stringent ban on reactor size would also further weaken the rationale for the U.S. ASAT system. The only Soviet war-fighting satellites that the Department of Defense has specifically identified as potential targets for this device are the nuclear-reactor-powered Soviet radar ocean reconnaissance satellites that can pinpoint large American ships for attack by long-range missiles. Verifying limits on nuclear reactors in space would be relatively easy because nuclear reactors in space produce waste heat that is radiated into space by vanes that glow dull red and are easily detected.

Fortunately, the arms control problems created by the U.S. space shuttle have largely faded. During the previous round of talks the Soviets insisted that the shuttle be defined as an ASAT system, because its ability to maneuver in orbit and to bring objects back to earth suggested an ability to destroy or hijack Soviet satellites. Yet the shuttle itself is not a very plausible ASAT weapon. And perhaps realizing that no space shuttle ban is likely, Soviet U.N. delegates have recently offered a draft treaty calling for bans on weapons carried aboard manned spacecraft, rather than on the craft themselves.

The Reagan administration and defense analysts outside the administration such as Colin Gray have stated that the prospective space treaties are not verifiable. Yet the United States maintains and is currently upgrading an elaborate system of space radars and advanced optical telescopes designed to track objects in near orbit several square inches in size and eventually to observe football-sized objects in geosynchronous orbit. Computers at the North American Aerospace Defense Command headquarters will soon be able to track thousands of objects at one time. There is little doubt that the United States is able to verify a soundly drafted ASAT space weapons ban.

To provide further protection against residual ASAT capabilities or some worst-case breakout scenario, satellites can be made more survivable. Survivability measures include: improving satellite maneuverability, increasing their numbers, insulating them from radiation effects, keeping in-space or ready-to-launch spares, and maintaining higher orbits. The United States has begun a fairly vigorous program to improve satellite survivability, reversing the earlier preference for a small number of large, vulnerable satellites. Given the inherent economies of destruction for an attacker, satellite survivability measures are not a substitute for ASAT arms control. But by raising the performance level an ASAT must meet to be useful, satellite survivability could enhance the political appeal of controls and diminish the incentives to cheat.

ASAT treaty skeptics argue that the Soviets have no real incentive to negotiate because Washington depends on the space assets jeopardized by ASATs more than does Moscow. Although the United States uses near space for strategic surveillance to a far greater degree than the Soviets, this dependence has little relevance to wartime uses of space. In fact, the Soviets may actually be more dependent on space assets for operations beyond their own borders, for they lack a first-class, far-flung system of bases and labor under a more rigid command philosophy.

However, time is running out. Once thoroughly tested, these systems will be a verification nightmare since many orbiting Soviet vehicles or U.S. fighter planes could harbor satellite killers. Unfortunately, the United States recently decided to accelerate testing the direct-ascent homing missile and to delay ASAT negotiations. Unilateral U.S. restraint, however, and an immediate resumption of negotiations would do more than further tests to enhance the security of both superpowers.

Linking Overarmed Societies

Outer space will never again be the geopolitical vacuum it once was. To make sure civilian space programs do not surreptitiously spawn weapons capability, a positive space security agenda should be pursued. Controlling space weapons can clear the stage for the use of space technology to improve both arms control verification and superpower relations. The most important items on a space security agenda are an improved Law of Space regime, an international satellite monitoring agency, and joint U.S.-Soviet space ventures. This last item, now perceived as irrelevant or at best symbolic of other changes, holds the greatest promise for defusing superpower conflict. Indeed, cooperation can link these overarmed societies in the pursuit of goals that transcend national differences, thus making the thorny verification question and breakout and peace conversion problems more tractable.

The two most critical space law issues are traffic control and the boundary between national airspace and international outer space. Thousands of tracked objects and tens of thousands of smaller untracked ones currently whiz around the earth. By reducing the chances of accidental collisions, close passes, and explosions, space traffic control will strengthen comprehensive weapons test bans.

Neither the Outer Space Treaty of 1968 nor nature draws a line dividing the atmosphere from near space. So far the absence of an agreed-upon upper limit of national airspace has been academic, since the highest-flying operational aircraft climbs only as high as 110,000 feet while the lowest satellite passes at 400,000 feet. But this convenient buffer zone will disappear, as aerospace planes such as the U.S. space shuttle or the proposed space cruiser begin to operate in both orbital space and the atmosphere. As the ability to shoot down a large, low-flying satellite spreads with the diffusion of air defense missiles, sounding rockets, and intermediate-range ballistic missiles, the United States could face a stratospheric incident with a regional power like that involving U.S. and Libyan fighter planes over the Gulf of Sidra in 1981. Rather than be caught off guard by such a crisis—another U-2 affair—or by a Soviet campaign to close the Open Skies by significantly raising the boundary of national space, the United States should recommend the lowest boundary possible.

Second, the superpowers should construct an internationally managed satellite reconnaissance and surveillance system. The president of War Control Planners, Inc., Howard Kurtz, has long promoted a multinational network of satellites, and the idea gained official support in 1978 when France proposed an International Satellite Monitoring Agnecy (ISMA). As outlined by then President Valéry Giscard d'Estaing before the U.N. General Assembly, the agency would extend the benefits of surveillance satellites to countries without space capability, would permit the U.N. Security Council to monitor crises and border disputes, and would verify the treaties banning chemical and biological warfare, as well as environmental modification for military purposes. Depending on whether ISMA obtained technology from the superpowers, a basic monitoring system would cost between $1 billion and $2 billion a year—more than the entire U.N. budget. The United States has vigorously opposed ISMA, arguing that it would be impossible for an agency operating by majority rule to handle sensitive issues of data interpretation. Support for ISMA has been strong among the many countries not likely to join the space club any time soon. The first question to resolve is who would have access to the satellite-generated data.

The superpowers' strong opposition to ISMA reflects a desire to continue monopolizing satellite technology. The U.S. position is, however, extemely shortsighted since the United States stands to lose most from a closing of the Open Skies. Unless satellite technology becomes more accessible, those countries excluded from its benefits may see it as a form of spying and favor a treaty that bans or limits such systems. Although ISMA's opponents protest its cost, if satellite verification of regional arms control efforts is half as successful as it has been between the superpowers, the agency could pay for itself many times over in reduced arms expenditures.

The greatest adventure of our time—the peaceful exploration of the cosmos—faces a bleak future because of declining budgets and growing monopolization of technology by the Soviet and U.S. militaries. Two separate space programs doing many of the same things make less and less sense—especially as the costs of missions rise. Joint planetary and scientific probes would boost both countries' sagging programs and improve political relations.

Joint manned missions would be even more valuable. A U.S. administration, looking for a bold, highly visible way to ease superpower tensions quickly, might find a joint U.S.-Soviet manned space mission particularly appealing. While the *Apollo-Soyuz* rendezvous in 1976 was a meeting of two parallel systems, the space shuttle and the *Salyut* stations complement each other perfectly. What better way to learn more about Soviet space stations or to allay Soviet suspicions about the shuttle than to take each other's astronauts and cosmonauts for a ride? More routine cooperation in space and the consequent mingling of space scientists would make concealed weapons work all but impossible to keep secret.

These positive security initiatives have not received the attention they deserve. But the Soviets take space exploration very seriously and herald it as one of the few irrefutable accomplishments of the October Revolution. And despite relative official neglect in recent years, the high frontier casts a powerful spell on the American psyche. Significant space cooperation could develop the political constituency and momentum in both countries that more traditional arms control efforts have sadly lacked.

Cooperating in Outer Space

A militarily neutral cosmos and the use of space technology to defuse superpower competition will not, however, end once and for all the threat posed to mankind by space technology. At the risk of oversimplifying, two rather stark alternative scenarios seem most plausible. One path involves large-scale space industrialization through solar-power satellites, large, free-floating space colonies, and asteroid and lunar mining. Another involves a more information-oriented path that makes use of near space to study remote regions of the universe and the Earth and to explore and colonize distant bodies. While the first set of options is probably infeasible, both economically and environmentally, such a path has strong appeal to the U.S. and Soviet space communities, which are still committed to megaengineering projects.

The first strategy of industrializing near space has an inescapable military dimension and contains the seeds of a planetary tyranny. The difference between civilian function and military capability has always been small in space—and it will all but disappear with the advent of large-scale space ventures such as solar-power satellites, space colonies, and asteroid mining. Solar-power satellites, first conceived for military purposes, may generate enough power to turn particle-beam and laser systems into viable weapons. And lunar or asteroidal mass accelerators—the slingshot-like devices to hurl construction material into orbit that many large-scale space industrialization schemes are based on—have a powerful destructive potential. A 1,000-ton lunar rock, 15 meters in diameter, hitting Earth at 12 to 20 kilometers per second, would release energy equivalent to a two-megaton nuclear blast. Unlike particle beams or lasers, which rapidly dissipate in the atmosphere, such slingshots would make all points on Earth vulnerable. This benign material transportation can be transformed into a weapon for mass destruction with a simple trajectory adjustment. These problems seem comfortably distant, yet a recent National Aeronautics and Space Administration conference on diverting asteroids whose orbits cross with Earth's recalled the military's long-standing ambition to learn how to bombard Earth with such objects.

Controlling the military potential of these technologies will be virtually impossible if the industrialization of space occurs in an environment of military competition. But their military use will be very real even if pursued in a cooperative, global, peaceful regime. The fundamental danger of large-scale space ventures rests on one inescapable geopolitical fact: effective control of outer space means effective control of the planet. Space has thus far been exploited as a vantage point. Should this high ground be garrisoned or colonized, a high, impregnable castle will have been created. Given the long and dreary history of coup d'états on Earth, how long will the denizens of these high castles, suffused with the elitist messianic spirit already so visible in the space community, wait before exploiting their *übermensch* status to set forth a new agenda for the species?

In contrast, a future devoted to learning more about the cosmos and dispersing humans more widely within it would minimize the threats of planetary tyranny inherent in our closed, crowded world. Because distances in space are so enormous, a scattering of the species would make local self-rule a necessity and centralized tyranny an impossibility, as in the days when seagoing ships only loosely connected far-flung societies. A knowledge- and exploration-oriented space future might also prevent a catastrophic collapse of the Earth's life-support system, caused by excessive industrialization, from extinguishing the human species.

Mankind already lives in a world decisively shaped by past choice about the use of space technology. As long as hostile nation-states seeking unilateral military advantage continue to make the basic decisions about these planet-embracing technologies, these decisions are unlikely to be sound. With space technology the human species is playing with a fire of awesome potential for destruction and creation. Unless these technologies are used to secure the whole Earth, they may flare disastrously out of control.

6 Finding Effective Lower-Risk Means of Defense

As many commencement speakers have said, "More than at any other time in history, humanity faces a crossroads." It was Woody Allen's comic genius to imagine such a talk continuing as follows: "One path leads to despair and utter hopelessness; the other, to total extinction. I pray we have the wisdom to choose wisely."

In looking at the arms race, many share this fatalism. The more frightened we get, the more we want to turn away from reality; and the more we suppress or deny, the worse the situation is allowed to become—the more we have to fear. When Franklin Roosevelt said, "All we have to fear is fear itself," he could not have meant there were no other dangers in the world. He meant that a nation is lost if it becomes paralyzed by fear—or, we might add, narcotized by denial. How can we learn to face reality, to turn our face toward it—that sensitive and vulnerable part—and then keep our wits in spite of what we see? In short, how do we avoid what Don Carlson calls "numbing out or dumbing out"?

In some eyes, reality today is very disturbing. For example, Edward Luttwak analyzes the Soviet regime in terms of its optimism and pessimism in two areas: its long-term future and its immediate military capacity relative to that of its adversaries. Under Khrushchev, the regime believed that its economic growth rate and ideological appeal would soon allow it to surpass the West. It would attend capitalism's funeral. On the other hand, despite Jack Kennedy's fear of a "missile gap," Soviet retaliatory and other nuclear forces were not overwhelming then, as compared with those of the U.S. and its allies.

By now, in contrast, the Soviets have created a menacing military establishment in almost every category. However, for many reasons analyzed by Luttwak, the regime feels pessimistic about its ability to compete with the West in other ways, especially over the long run. In this situation, the Soviets may feel tempted to use their brute power while they have it. Luttwak warns that if the nations of Western Europe and Japan "simply refuse to respond to anything short of a direct attack, thus undermining both the capacity and incentive of an American response, the Soviet Union will be set on the road to war—a war neither Western nor nuclear but quite possibly catastrophic all the same."[1]

From this analysis, Luttwak draws the conclusion that we need to develop a much more capable military than we now have—another subject to which he has usefully turned his mind, as a group of Senators has also begun to do. Certainly, as long as the U.S. and Soviet Union are playing the grand game described by Luttwak, his prescriptions are a valuable corrective to wishful assumptions and slack thought. A continuation of this dangerous game, until an abrupt disastrous ending, is arguably our most likely future. In this book and its companion, however, we explore the possibility for changing the nature of the game.[2]

If the strategic "moves" described by Luttwak were the behavior not of a nation but of an individual, a therapist would probably talk about "displacement activity." If an activity based upon a real need is frustrated, an organism turns to a substitute. For example, if the need is sexual, the substitute might be eating or fighting. A complex society differs from an individual in many ways, but let's say that its basic needs, apart from material supplies, are security and self-esteem—to feel safe and to feel good about itself. Say further that, as so often happens, the society decides it won't be safe unless it terrifies anybody who might otherwise attack. In that case, the society substitutes the activity of threatening possible enemies for the activity of seeking security in other ways. It may forget there *are* other ways, especially if those ways are deeply violated by the activity of terrifying people.

Except at its highest, civilization consists of displacement activities, which, at best, are incapable of providing the satisfactions we really want. At worst, they are dangerous, self-defeating, costly, and confusing—a pathology of defense. The nuclear arms race is a case in point. Even the "realists" ask, as Kissinger belatedly did: at this level of numbers, what in God's name *is* strategic superiority? How do we make use of the ability to destroy civilization?

Finding Effective Lower-Risk Means of Defense

There are alternatives.[3] In this section, the Delhi Declaration asserts a role for the rest of the world in the superpower quarrel. Kennan calls for deep cuts in nuclear stockpiles—half, for starters, applying to every category of nuclear arms, with compliance verified by the intelligence agencies of each side. Salter discusses a way to make more complex cuts, based upon an ingenious adaptation of the children's rule in dividing a piece of cake: "I cut, you choose." Dyson discusses how some of the big bullying weapons can perhaps be defeated by small smart defenses, with the help of microcomputers. Sommer explores the concept of "non-provocative defense"—a military approach overshadowed today by the threat to retaliate in kind. Sharp goes beyond armaments to discuss non-military kinds of defense. All of them keep clear that our goal is not to terrify other people; it's to be secure.

—*Craig Comstock*

1. Edward N. Luttwak, *The Grand Strategy of the Soviet Union* (St. Martin's Press, 1983), with appendices on the economy by Herbert Block and on the military by W. Seth Carus.
2. The companion volume is *Citizen Summitry: Keeping the Peace When It Matters Too Much to be Left to Politicians,* edited by Don Carlson and Craig Comstock (Tarcher/St. Martin's Press, 1986).
3. For example, see Dietrich Fischer, *Preventing War in the Nuclear Age* (Rowman & Allanheld, 1984); John Galtung, *There Are Alternatives! Four Roads to Peace and Security* (Spokesman/DuFour Editions, 1984); Gene Sharp, *Making Europe Unconquerable: The Potential of Civilian-Based Deterrence and Defense* (Ballinger, 1985); and Mark Sommer (for the Exploratory Project on the Conditions of Peace), *Beyond the Bomb: Living Without Nuclear Weapons* (Expro Press/The Talman Company, 1985).

The Delhi Declaration is less remarkable for exactly what it says than for who is saying it. Signed on January 28, 1985, the brief document notes that "a small group of men and machines in cities far away . . . decide our fate." It warns about overkill and nuclear winter. It pretends to believe that the U.S. and Soviet Union seriously want to "prevent an arms race in space and terminate it on earth, ultimately to eliminate nuclear arms everywhere." It asks for an end to all testing of nuclear weapons, culminating in a treaty. This would be followed by "substantial reductions in nuclear forces," which would allow the transfer of resources to social and economic development. The declaration promises to renew the appeal until the nuclear weapons states take notice.

None of these points would surprise any observer of arms control efforts. What's striking are the context in which the declaration was drafted and the signatories. They include Raul Alfonsin of Argentina, Miguel de la Madrid of Mexico, and Julius Nyerere of the United Republic of Tanzania—each is President of his country—Prime Minister Rajiv Gandhi of India, and his counterparts Andreas Papandreou of Greece and the late Olof Palme of Sweden. Drafting of the declaration was initiated by a group now called Parliamentarians Global Action—600 legislators from thirty countries who believe that "the only way to provide security for their constituents, and to ensure a future for humanity, is through comprehensive disarmament and fundamental reform of the international system."

Many declarations are signed by considerably more than six people. However, very few people who sign petitions are able, as these six are, to make state visits to talk with the President of the U.S. or the General Secretary of the Communist Party of the U.S.S.R. And yet, though each of the signatories heads a substantial nation, they share a sense of being outside the superpower club, outside the circle that decides the fate of humanity. In this way, they stand for the rest of us.

Delhi Declaration

Five Continent Peace Initiative

FORTY YEARS AGO, WHEN ATOMIC BOMBS WERE BLASTED over Hiroshima and Nagasaki, the human race became aware that it could destroy itself, and horror came to dwell among us. Forty years ago, also, the nations of the world gathered to organise the international community, and with the United Nations hope was born for all people.

Finding Effective Lower-Risk Means of Defense

Almost imperceptibly, over the last four decades, every nation and every human being has lost ultimate control over their own life and death. For all of us, it is a small group of men and machines in cities far away who can decide our fate. Every day we remain alive is a day of grace as if mankind as a whole were a prisoner in the death cell awaiting the uncertain moment of execution. And like every innocent defendant, we refuse to believe that the execution will ever take place.

We find ourselves in this situation because the nuclear weapon states have applied traditional doctrines of war in a world where new weapons have made them obsolete. What is the point of nuclear "superiority" or "balance" when each side already has enough weapons to devastate the earth dozens of times over? If the old doctrines are applied in the future, the holocaust will be inescapable sooner or later. But nuclear war can be prevented if our voices are joined in a universal demand in defence of our right to live.

As a result of recent atmospheric and biological studies, there have been new findings which indicate that in addition to blast, heat and radiation, nuclear war, even on a limited scale, would trigger an arctic nuclear winter which may transform the Earth into a darkened, frozen planet posing unprecedented peril to all nations, even those far removed from the nuclear explosions. We are convinced that this makes it still more pressing to take preventive action to exclude forever the use of nuclear weapons and the occurrence of a nuclear war.

In our joint statement of May 22, 1984, we called upon the nuclear weapon states to bring their arms race toa halt. We are encouraged by the world-wide response to our appeal. The international support we received, and the responses of the nuclear weapon states themselves, have been such that we deemed it our duty to meet here in New Delhi to consider ways to further our efforts.

The nuclear weapons states have a particular responsibility for the dangerous state of the arms race. We urge them to join us in the search for a new direction. We welcome the agreement in Geneva on January 8, 1985, between the Soviet Union and the United States to start bilateral negotiations on "a complex of questions concerning space and nuclear arms—both strategic and intermediate range—with all the questions considered and resolved in their inter-relationship." We attach great importance to the proclaimed objective of these negotiations: to prevent an arms race in space and to terminate it on earth, ultimately to eliminate nuclear arms everywhere. We expect the two major nuclear weapon powers to implement, in good faith, their undertaking and their negotiations to produce, at an early date, significant results. We will follow their work closely and we expect that they will keep the international community informed of its progress. We stress that the agenda for and the outcome of these negotiations is a matter of concern for all nations and all people.

We reiterate our appeal for an all-embracing halt to the testing, production and deployment of nuclear weapons and their delivery systems. Such a halt would greatly facilitate negotiations. Two specific steps today require special attention: the prevention of an arms race in outer space, and a comprehensive test ban treaty.

Outer space must be used for the benefit of mankind as a whole, not as a battleground of the future. We therefore call for the prohibition of the development, testing, production, deployment and use of all space weapons. An arms race in space would be enormously costly, and have grave destabilising effects. It would also endanger a number of arms limitation and disarmament agreemtns.

We further urge the nuclear weapon states to immediately halt the testing of all kinds of nuclear weapons, and to conclude, at an early date, a treaty on a nuclear weapon test ban. Such a treaty would be a major step towards ending the continuous modernisation of nuclear arsenals.

We are convinced that all such steps, in so far as necessary, can be accompanied by adequate and non-discriminatory measures of verification.

A halt to the nuclear arms race is at the present moment imperative. Only thus can it be ensured that nuclear arsenals do not grow while negotiations proceed. however, this halt should not be an end in itself. It must be immediately followed by substantial reductions in nuclear forces, leading to the complete elimination of nuclear weapons and the final goal of General and Complete Disarmament. Parallel to this process, it is urgently necessary to transfer precious resources currently wasted in military expenditure to social and economic development. The strengthening of the United Nations must also be an essential part of this endeavour.

It is imperative to find a remedy to the existing situation where hundreds of billions of dollars, amounting to approximately one and a half million per minute, are spent annually on weapons. This stands in dramatic contrast to the poverty, and in some cases misery, in which two-thirds of the world population lives.

The future of all peoples is at stake. As representatives from non-nuclear weapon states, we will not cease to express our legitimate concern and make known our demands. We affirm our determination to facilitate agreement among the nuclear weapon states, so that the required steps can be taken. We will seek to work together with them for the common security of mankind and for peace.

We urge people, parliaments and governments the world over to lend forceful support to this appeal. Progress in disarmament can only be achieved with an informed public applying strong pressure on governments. Only then will governments summon the necessary political will to overcome the many obstacles which lie in the path of peace. The World Disarmament Campaign launched by the United Nations represents a very important element in generating that political will.

For centuries, men and women have fought for their rights and freedoms. We now face the greatest struggle of all—for the right to live, for ourselves and for future generations. . . .

George F. Kennan combines direct observation of the Soviet Union and of the American government, historical savvy, and the experience of diplomatic responsibility. Bold in argument yet circumspect in his treatment of evidence, deeply cultured, unblinking in his view of tyranny, imaginative in his sense of the future, Kennan is what the Japanese would call a "living national treasure."

From serving in Moscow during the time of Stalin, Kennan has long known in his bones that millions can suddenly be killed—as by a tyrant, so by the bomb. It can actually happen. Frustrated by prolonged arms control negotiations—which are not even about reducing the numbers of weapons, but simply about restraining further testing and deployment—he offers a simple proposal. If both sides would benefit from a deep cut in the number of weapons, let's not haggle over "asymmetries" in the "force posture" of the two sides. As a first step, let's agree to cut all categories of weapons by half, to be verified by the intelligence services of both sides. Then let's make another big cut.

Kennan knows that merely reducing the number of weapons will not save us. The current stockpiles are so grotesque that a tiny fraction of them could destroy our civilization. But making a deep cut would do what Don Carlson calls "breaking the trance." By reminding us that another way is possible, it would resuscitate our imaginations.

Deep Cuts

George F. Kennan

ADEQUATE WORDS ARE LACKING TO EXPRESS THE FULL seriousness of our present situation. It is not just that we are for the moment on a collision course politically with the Soviet Union, and that the process of rational communication between the two governments seems to have broken down completely; it is also—and even more importantly—the fact that the ultimate sanction behind the conflicting policies of these two governments is a type and volume of weaponry which could not possibly be used without utter disaster for us all.

Finding Effective Lower-Risk Means of Defense

For over thirty years, wise and far-seeing people have been warning us about the futility of any war fought with nuclear weapons and about the dangers involved in their cultivation. Some of the first of these voices to be raised were those of great scientists, including outstandingly that of Albert Einstein himself. But there has been no lack of others. Every president of this country, from Dwight Eisenhower to Jimmy Carter, has tried to remind us that there could be no such thing as victory in a war fought with such weapons. So have a great many other eminent persons.

When one looks back today over the history of these warnings, one has the impression that something has now been lost of the sense of urgency, the hopes, and the excitement that initially inspired them, so many years ago. One senses, even on the part of those who today most acutely perceive the problem and are inwardly most exercised about it, a certain discouragement, resignation, perhaps even despair, when it comes to the question of raising the subject again. The danger is so obvious. So much has already been said. What is to be gained by reiteration? What good would it now do?

Look at the record. Over all these years the competition in the development of nuclear weaponry has proceeded steadily, relentlessly, without the faintest regard for all these warning voices. We have gone on piling weapon upon weapon, missile upon missile, new levels of destructiveness upon old ones. We have done this helplessly, almost involuntarily: like victims of some sort of hypnosis, like men in a dream, like lemmings heading for the sea, like the children of Hamlin marching blindly along behind their Pied Piper. And the result is that today we have achieved, we and the Russians together, in the creation of these devices and their means of delivery, levels of redundancy of such grotesque dimensions as to defy rational understanding.

I say redundancy. I know of no better way to describe it. But actually, the word is too mild. It implies that there could be levels of these weapons that would not be redundant. Personally, I doubt that there could. I question whether these devices are really weapons at all. A true weapon is at best something with which you endeavor to affect the behavior of another society by influencing the minds, the calculations, the intentions, of the men that control it; it is not something with which you destroy indiscriminately the lives, the substance, the hopes, the culture, the civilization, of another people.

What a confession of intellectual poverty it would be—what a bankruptcy of intelligent statesmanship—if we had to admit that such blind, senseless acts of destruction were the best use we could make of what we have come to view as the leading elements of our military strength!

To my mind, the nuclear bomb is the most useless weapon ever invented. It can be employed to no rational purpose. It is not even an effective defense against itself. It is only something with which, in a moment of petulance or panic, you commit such fearful acts of destruction as no sane person would ever wish to have upon his conscience.

There are those who will agree, with a sigh, to much of what I have just said, but will point to the need for something called deterrence. This is, of course, a concept which attributes to others—to others who, like ourselves, were born of women, walk on two legs, and love their children, to human beings, in short—the most fiendish and inhuman of tendencies.

But all right: accepting for the sake of argument the profound iniquity of these adversaries, no one could deny, I think, that the present Soviet and American arsenals, presenting over a million times the destructive power of the Hiroshima bomb, are simply fantastically redundant to the purpose in question. If the same relative proportions were to be preserved, something well less than 20 percent of those stocks would surely suffice for the most sanguine concepts of deterrence, whether as between the two nuclear superpowers or with relation to any of those other governments that have been so ill-advised as to enter upon the nuclear path. Whatever their suspicions of each other, there can be no excuse on the part of these two governments for holding, poised against each other and poised in a sense against the whole Northern Hemisphere, quantities of these weapons so vastly in excess of any rational and demonstrable requirements.

How have we got ourselves into this dangerous mess?

Let us not confuse the question by blaming it all on our Soviet adversaries. They have, of course, their share of the blame, and not least in their cavalier dismissal of the Baruch Plan so many years ago. They too have made their mistakes; and I should be the last to deny it.

But we must remember that it has been we Americans who, at almost every step of the road, have taken the lead in the development of this sort of weaponry. It was we who first produced and tested such a device; we who were the first to raise its destructiveness to a new level with the hydrogen bomb; we who introduced the multiple warhead; we who have declined every proposal for the renunciation of the principle of "first use"; and we alone, so help us God, who have used the weapon in anger against others, and against tens of thousands of helpless noncombatants at that.

I know that reasons were offered for some of these things. I know that others might have taken this sort of lead, had we not done so. But let us not, in the face of this record, so lose ourselves in self-righteousness and hypocrisy as to forget our own measure of complicity in creating the situation we face today.

What is it then, if not our own will, and if not the supposed wickedness of our opponents, that has brought us to this pass?

The answer, I think, is clear. It is primarily the inner momentum, the independent momentum, of the weapons race itself—the compulsions that arise and take charge of great powers when they enter upon a competition with each other in the building up of major armaments of any sort.

This is nothing new. I am a diplomatic historian. I see this same phenomenon playing its fateful part in the relations among the great European powers as much as a century ago. I see this competitive buildup of armaments conceived initially as a means to an end but soon becoming the end itself. I see it taking possession of men's imagination and behavior, becoming a force in its own right, detaching itself from the political differences that initially inspired it, and then leading both parties, invariably and inexorably, to the war they no longer know how to avoid.

This is a species of fixation, brewed out of many components. There are fears, resentments, national pride, personal pride. There are misreadings of the adversary's intentions—sometimes even the refusal to consider them at all. There is the tendency of national communities to idealize themelves and to dehumanize the opponent. There is the blinkered, narrow vision of the professional military planner, and his tendency to make war inevitable by assuming its inevitability.

Tossed together, these components form a powerful brew. They guide the fears and the ambitions of men. They seize the policies of governments and whip them around like trees before the tempest.

Is it possible to break out of this charmed and vicious circle? It is sobering to recognize that no one, at least to my knowledge, has yet done so. But no one, for that matter, has ever been faced with such great catastrophe, such inalterable catastrophe, at the end of the line. Others, in earlier decades, could befuddle themselves with dreams of something called "victory." We, perhaps fortunately, are denied this seductive prospect. We have to break out of the circle. We have no other choice.

How are we to do it?

I must confess that I see no possibility of doing this by means of discussions along the lines of the negotiations that have been in progress, off and on, over this past decade, under the acronym of SALT. I regret, to be sure, that the most recent SALT agreement has not been ratified. I regret it, because if the benefits to be expected from that agreement were slight, its disadvantages were even slighter; and it had a symbolic value which should not have been so lightly sacrificed.

But I have, I repeat, no illusion that negotiations on the SALT pattern—negotiations, that is, in which each side is obsessed with the chimera of relative advantage and strives only to retain a maximum of the weaponry for itself while putting its opponent to the maximum disadvantage—I have no illusion that such negotiations could ever be adequate to get us out of this hole. They are not a way of escape from the weapons race; they are an integral part of it.

Whoever does not understand that when it comes to nuclear weapons the whole concept of relative advantage is illusory—whoever does not understand that when you are talking about absurd and preposterous quantities of overkill the relative sizes of arsenals have no serious meaning—whoever does not understand that the danger lies, not in the possibility that someone else might have more missiles and warheads than we do, but in the very existence of these unconscionable quantities of highly poisonous explosives, and their existence, above all, in hands as weak and shaky and undependable as those of ourselves or our adversaries or any other mere human beings: whoever does not understand these things is never going to guide us out of this increasingly dark and menacing forest of bewilderments into which we have all wandered.

I can see no way out of this dilemma other than by a bold and sweeping departure that would cut surgically through the exaggerated anxieties, the self-engendered nightmares, and the sophisticated mathematics of destruction in which we have all been entangled over these recent years, and would permit us to move, with courage and decision, to the heart of the problem.

President Reagan recently said, and I think very wisely, that he would "negotiate as long as necessary to reduce the numbers of nuclear weapons to a point where neither side threatens the survival of the other."

Now that is, of course, precisely the thought to which these present observations of mine are addressed. But I wonder whether the negotiations would really have to be at such great length. What I would like to see the president do, after due consultation with the Congress, would be to propose to the Soviet government an immediate across-the-boards reduction by 50 percent of the nuclear arsenals now being maintained by the two superpowers; a reduction affecting in equal measure all forms of the weapon, strategic, medium-range, and tactical, as well as all means of their delivery; all this to be implemented at once and without further wrangling among the experts, and to be subject to such national means of verification as now lie at the disposal of the two powers.

Whether the balance of reduction would be precisely even—whether it could be construed to favor statistically one side or the other—would not be the question. Once we start thinking that way, we would be back on the same old fateful track that has brought us where we are today. Whatever the precise results of such a reduction, there would still be plenty of overkill left—so much so that if this first operation were successful, I would then like to see a second one put in hand to rid us of at least two-thirds of what would be left.

Now I have, of course, no idea of the scientific aspects of such an operation; but I can imagine that serious problems might be presented by the task of removing, and disposing safely of, the radioactive contents of the many thousands of warheads that would have to be dismantled. Should this be the case, I would like to see the president couple his appeal for a 50 percent reduction with the proposal that there be established a joint Soviet-American scientific committee, under the chairmanship of a distinguished neutral figure, to study jointly and in all humility the problem not only of the safe disposal of these wastes but also of how they could be utilized in such a way as to make a positive contribution to human life, either in the two countries themselves or—perhaps preferably—elsewhere. In such a joint scientific venture we might both atone for some of our past follies and lay the foundation for a more constructive relationship.

It will be said this proposal, whatever its merits, deals with only a part of the problem. This is perfectly true. Behind it there would still lurk the serious political differences that now divide us from the Soviet government. Behind it would still lie the problems recently treated, and still to be treated, in the SALT forum. Behind it would still lie the great question of the acceptability of war itself, any war, even a conventional one, as a means of solving problems among great industrial powers in this age of high technology.

What has been suggested here would not prejudice the continued treatment of these questions just as they might be treated today, in whatever forums and under whatever safeguards the two powers find necessary. The conflicts and arguments over these questions could all still proceed to the heart's content of all those who view them with such passionate commitment. The stakes would simply be smaller; and that would be a great relief to all of us.

What I have suggested is, or course, only a beginning. But a beginning has to be made somewhere; and if it has to be made, is it not best that it should be made where the dangers are the greatest, and their necessity the least? If a step of this nature could be successfully taken, people might find the heart to tackle with greater confidence and determination the many problems that would still remain.

It will also be argued that there would be risks involved. Possibly so. I do not see them. I do not deny the possibility. But if there are, so what? Is it possible to conceive of any dangers greater than those that lie at the end of the collision course on which we are now embarked? And if not, why choose the greater—why choose, in fact, the greatest—of all risks, in the hopes of avoiding the lesser ones?

We are confronted here, my friends, with two courses. At the end of the one lies hope—faint hope, if you will, uncertain hope, hope surrounded with dangers, if you insist. At the end of the other lies, so far as I am able to see, no hope at all.

Can there be—in the light of our duty not just to ourselves (for we are all going to die sooner or later) but of our duty to our own kind, our duty to the continuity of the generations, our duty to the great experiment of civilized life on this rare and rich and marvelous planet—can there be, in the light of these claims on our loyalty, any question as to which course we should adopt?

In the final week of his life, Albert Einstein signed the last of the collective appeals against the devlopment of nuclear weapons that he was ever to sign. He was dead before it appeared. It was an appeal drafted, I gather, by Bertrand Russell. I had my differences with Russell at the time as I do now in retrospect; but I would like to quote one sentence from the final paragraph of that statement, not only because it was the last one Einstein ever signed, but because it sums up, I think, all that I have to say on the subject. It reads as follows: "We appeal, as human beings to human beings: Remember your humanity, and forget the rest."

If superpowers had similar geographical positions and if each of them had developed exactly the same weapons, in the same quantities, agreements to reduce weaponry would be less difficult than they have so far proven to be. In fact, for such reasons as geopolitics, the sequence of technological discovery, military tradition, and bureaucratic struggle, the superpowers have a quite different mix of weapons. Thus, there is no obvious way to compare the value of a given American weapon with that of a Soviet weapon.

In arms control negotiations, each side generally proposes to limit the weapons it most fears, while permitting the ones in which it enjoys a relative strength. Deterrence requires that the enemy know about the most frightening weapons, but in the game of arms control each side complains that the other side is ahead and ought to give up some of its especially sinister weapons. When the other side declines to agree, it's charged with insincerity. Each side seeks to acquire "bargaining chips" in the form of new weapons, whether or not they are militarily useful or destabilizing; but once these chips are on the table, they are generally regarded as too valuable to give away. Both sides play this game. What else would you expect?

If two governments really wanted to cut back, each could agree to place a relative value on each weapon that it deploys, with the expectation of cutting back the stockpile by, say, five percent of the total value. It's the genius of Stephen Salter's plan that side A would be permitted to pick which weapons Side B would dismantle and vice versa, based upon the valuations assigned by their owners. This would serve as an incentive to evaluate wisely and as an assurance that both sides would regard the transaction as fair.

Once we adopt the principle of what Fisher and Ury call "negotiation on the merits," rather than "positional bargaining," many ingenious methods come to mind. As Salter points out, his method is based upon the practice of one child cutting the piece of cake to be shared, and the other having first choice. This doesn't tax the limits of game theory, but it might deal with some of the sticking points in arms reduction. Are there other childhood games we could learn from?

I Cut, You Choose

Stephen H. Salter

DESPITE THE STRONG WISHES OF A LARGE MAJORITY OF the world's people, progress in reaching international agreement on the reduction of nuclear weapons has been very slight.[1] Despite the work of the cleverest minds and the expenditure of ever-increasing amounts of money, world security seems more and more remote. Disarmament issues appear more and more complicated, and the arguments about how to reduce nuclear weapons grow into head-spinning iterations of "we think that they think that we think that they think." While some experts are close to despair, others argue that the problems are too difficult for us outsiders to understand and that the solutions have to be left to them—the successors of the people who began it all.

If we are to emerge from this dangerous situation, we must understand how we got into it. We can begin with an analogy developed by Samuel Gorovitz, professor of philosophy at the University of Maryland.[2]

Gorovitz describes an auction with a difference. The goods will go to the highest bidder as in a normal auction, but the second-highest bidder will also have to pay whatever he bids yet get nothing in return. Let us suppose that the article for sale is a dollar bill and that the auctioneer invites an opening bid of 10 cents. At this point the profit margin appears very attractive, and plenty of people will be found willing to make a small bid for such a handsome return. Unfortunately, there will be others willing to raise the bid to 15 cents. While it is splendid to buy a dollar bill for only 10 cents, it is not good to pay 10 cents and get nothing. The sensible step for the opening bidder is clearly to raise the bid to 20 cents.

The two bidders are now locked into the trap devised by the rules of the auction. As the bidding proceeds their possible profit steadily declines, and the penalty for coming in second steadily rises. A smile appears on the face of the auctioneer—and, I dare say, the forked tail concealed by his rostrum twitches with delight—as the bidding passes 50 cents. The smile gets wider as each victim tries to force the other into second place and the bids get bigger. In psychological experiments it has been shown that people will bid up to five times the value of the article that they were originally tempted to buy for a tenth of its value—a 50-to-1 change. The auctioneer can win 10 times his risk capital.

We should note that it is only the rivalry between the bidders that gives the auctioneer his chance at profit. If the bidders could have been persuaded to cooperate by making a joint single bid of 10 cents, they could have shared the dollar and put the auctioneer out of business. In psychological experiments subjects have learned to foil the auctioneer by cooperating, but if rival bidders insisted dogmatically on outbidding each other, the competition would be endless.

Despite its simplicity the Gorovitz auction is an acutely accurate model of the progress of an arms race. It shows how two rivals can move to a dangerous and irrational situation through a series of apparently sensible steps. In 1945 it seemed as though nuclear weapons would provide lasting peace at low cost. There need be no more military service or dead young men. The initial bid was low. A very few bombs would do.

However, neither superpower could allow the other to make the higher bid. Each increase had to be matched, as in the Gorovitz auction. Inevitably, as time has gone on, the size of the nuclear weapon bids has risen. The superpowers keep hoping that each new one will provide the longed-for security, but each in turn proves to be inadequate. It is most interesting to retrieve from the archives the proposals for yesterday's weapons, now deemed insufficient. We can safely predict that the increasingly expensive weapons proposed for tomorrow will be deemed insufficient the day after tomorrow.

In the Gorovitz auction the bidders can escape their precarious position through cooperation. During the 40 years of the nuclear arms race, however, a number of powerful barriers to cooperation between the superpowers have been erected. These include:

Mutual suspicion
The excessive size of proposed weapons reductions
Loopholes in previous arms control agreements
The difficulties of weapon comparison
The perception of unequal starting points
Anxieties about verification

A close look at these barriers indicates that some could be reduced and others could actually be turned to advantage by an unconventional approach to reducing nuclear arms.

Suspicion. Suspicion between the two sides is very deep. The original reasons for it have been overlaid by the nuclear arms race itself. Each side used to make nuclear weapons because it hated the other's social system. Now it makes nuclear weapons because the other makes nuclear weapons. Any suggestion made by one side to reduce nuclear weapons is rejected instantly. The fact that Side A proposes something is taken as quite sufficient evidence that it would be to the disadvantage of Side B.

To overcome such suspicion a disarmament process should be totally and obviously symmetrical. The need for extended negotiations and the exchange of contentious proposals should be minimized.

Oversized steps. Proposals for sweeping arms reductions have a dramatic quality that makes them attractive to the public. However, it is reasonable to expect that the anxieties about accepting a reduction will be proportional to its magnitude. It should be easier to risk a very small step, since there would be time for the superpower to weigh its security position after taking the step and to back out before its security were threatened.

A series of very small reductions is more likely to be accepted than a total ban or immediate large reductions. Moreover, an actual reduction, no matter how small, would produce a relaxation of tension much greater than its military significance.

Loopholes. Disarmament agreements that prohibit particular types of weapons lead to frantic developments in the areas that escape the ban. If there are limits on the number of launchers, we get multiple warheads. If there are limits on throw weight, we get miniaturization, high-yield explosives, and terminal guidance. If long-range strategic weapons are stopped, then there will be forward-based, short-range tactical ones. If quantity is curtailed, quality will improve. It is therefore not clear that the results of overspecified arms restraints leave the participants any better off. I conclude that for a disarmament system to succeed, weapon definitions should be as plain and as broad as possible.

Weapon comparison. The complex characteristics of nuclear weapons are difficult to measure, thereby bedeviling attempts to compare them. Negotiators must weigh the destructive power of the warheads, the payload of the launchers, the time required for launch preparation, the flight time to the target, the accuracy of the guidance systems, the sophistication of electronic countermeasures, the flexibility of target alteration, the mobility and detectability of carrier vehicles, the hardness of the silos, the bandwidth and immunity of communication networks, the radar cross-section of aircraft and missiles, and the sonar signature of submarines. Every judgment will change in response to technical advances on either side. Comparisons would be difficult enough in calm debate among the officials of one side equipped with accurate knowledge of their own weapons. When the uncertainty of information gained by espionage is combined with hostile international relationships, the problem becomes intractable.

I therefore take it as axiomatic that disarmament proposals that rely on the two sides agreeing on exact weapon comparisons will fail. It is, however, possible to turn the difficulty to advantage and design a procedure that works better because of differences of opinion.

Unequal starting points. The two sides claim that disarmament may be possible if only their starting points were the same. By concentrating on particular categories of weapons deployed in particular areas, each side can argue that the other has or is building a dangerous advantage. Each overlooks—perhaps by intention—its own superiority in other categories of weapons deployed elsewhere.

Because each side claims to be inferior to its opponent and rejects any route to disarmament that widens the gap, we need a mechanism that magnifies the importance of the reduction made by each side in the eyes of its opponent. More specifically, if we have disarmament negotiations between two hostile superpowers—A and B—we need a disarmament mechanism in which A's reductions look bigger to B than they do to A, while B's reductions look bigger to A than they to do B.

Such an apparently paradoxical mechanism can be created, but it requires that we draw inspiration from what seems an unlikely source—the way that children divide a piece of cake.

A child's absolute happiness is usually not much affected by whether or not there is cake for tea, but children are acutely sensitive to relative benefits. If there is cake for tea and one child thinks another is getting more, distress results. Children have devised an elegant solution: The I-cut-you-choose rule. The chooser tries to detect any minor imprecision in the cut and turn it to slight advantage. The cutter tries to minimize this by being as accurate as possible.

Mathematicians like cake as much as anyone else.[3] They have been able to demonstrate the intriguing fact that if there are irregularities in the cake, such as more icing in one place or more cherries in another, and if the sharers differ about the relative values of icing and cherries, *both* parties to the division can feel sure that they have come off better. This assurance is exactly what we need for nuclear disarmament.

A cake-sharing disarmament agreement would work as follows. Each side first writes out a list of all the weapons in its own nuclear inventory. To each weapon it assigns a number, which I shall call a *military value percentage*. The sum of all the military value percentages is made equal to 100. The military value percentage represents the military usefulness of the weapon as perceived by its owner. If one side possesses 25,000 warheads of equal usefulness, for example, the military value percentage of each warhead will be 0.004. Because weapons vary a great deal, some will be assigned a higher value than 0.004 and some a lower value.

The numbers do not represent any sort of absolute value, because as mentioned earlier no such mutually agreed-upon value could be determined. The numbers are purely the relative values of the weapons in the eyes of their owners. What matters is that neither side should be able to claim that the loss of one small part of its inventory would reduce its security more than the loss of a different small part with an equal military value percentage.

Indeed, it would be almost impossible for the two sides to agree on the value of weapons, because the feeling of threat induced by every weapon in its potential victim rarely, if ever, matches the feeling of security that the same weapon provides to its owner. For example, a mobile, forward-based, quick launch unhardened missile with accurate terminal guidance can very easily be seen by its potential target as a serious first-strike threat whereas its owner thinks its contribution to security is minimal. However, a less accurate submarine-based weapon that is safe from detection and could be used at leisure for retaliation as a second-strike weapon offers a splendid protection for its owner without appearing too threatening to the target. Therefore, when each side in the negotiation looks at the other's list, it sees some weapons that have been assigned military value percentages lower than the threat it perceives from that weapon. Differences of opinion about weapon valuation have proved the stumbling block for other negotiation plans. The cake-sharing method uses such differences to provide the incentive to negotiate.

The critical step in applying the cake-sharing method is that each side picks weapons adding up to some agreed small total military value percentage—perhaps 1—from the list of its opponent and asks that these be dismantled in return for the same percentage of its own list chosen by the other side. Each side, of course, has to pretend to be quite indifferent to which of its own weapons are selected by its adversary, because the numbers were supposedly chosen to make all possible choices equal.

It is quite clear that because there is very likely to be a difference of opinion about each weapon as a source of security and threat, both sides can feel that they have obtained an advantage. Each believes it has removed the nastiest-looking devices threatening it and has paid for this pleasure with a standardized reduction of its own forces. Thus, the perceived gap between the two sides will have been narrowed. Side A will think that its own inventory is now 99 percent of what it had originally. A will also think that Side B's new inventory is less than 99 percent of Side B's original inventory—because Side A perceives that the weapons given up by Side B are more threatening than the military percentage value assigned to them by Side B. As a result the total number of weapons in the world will have gone down, and the process will have selectively reduced weapons that have a high threat value while leaving behind those with a high security value. Only a side that secretly knows it has supremacy and is determined to retain it could have any logical reservations.

After the first reduction, both sides should pause to reconsider the military value percentages in the light of their remaining inventory and the new knowledge of the opponent's views. Each side must feel quite confident that it cannot be forced into being left with a mix of weapons insufficient to protect its security. The chance for adjustment between every round prevents one side from making excessive selections from one section of its opponent's list. If one side does so, its opponent can raise the military value percentages of its remaining weapons so that fewer of them would be lost at each reduction. Balance is therefore preserved. Indeed, it is possible that by judicious choice of military value percentages, a side that began the process with a poor balance among its various weapons could move toward a better one. A slow, steady series of microsteps that can be steered by both sides is much less risky than rapid transitions.[4]

Cake-sharing reductions can provide an interesting channel of communication between the participating sides. Both the numbers in the military value percentage lists and the types of weapons selected for reduction convey much about intentions. It is possible to write "friendly" lists of military value percentages that put high values on one's own second-strike deterrent systems and low figures on first-strike systems. This should entice one's opponent into continuing the procedure. Antagonism is possible but unlikely to emerge when lists are revised. After one side analyzes its opponent's choices, it can rewrite its own list so that no particular choice looks especially attractive to the opponent.

Verification. Cake sharing has two features that should also minimize the problem of verification, our last barrier to cooperation in reducing nuclear arms. First, it is reasonable to expect that the larger the disarmament step, the greater the anxiety about cheating and the larger the military catastrophe if one side feels cheated. However, cheating raises much less concern when the steps are so small that they do not significantly affect the overall balance. If a cheating claim is made during the cake-sharing negotiations, futher reductions can be stopped until the matter is resolved.

Second, we can assume that the probability of detecting violations will rise with the time available for detection. There will be more satellite passes and opportunities for espionage. Intelligence reports can be digested more thoroughly and correlated more widely. The slowness of the cake-sharing method minimizes the serious difficulties of verification.

New approaches to verification can help. Some of the money saved through weapon reductions should be spent on surveillance satellites, for example. It may also be possible to allow the other side partial access to the computer systems that record the movements and service logs of individual military units. The data could be handed over after a delay sufficiently long to remove tactical military risks and then scrutinized in the manner of tax returns. A number of technically advanced, neutral countries would probably be willing to provide teams for intrusive inspection visits. Finally, new technological developments are improving the capabilities of nonintrusive monitoring instruments.

History should encourage us: only a few accusations of breaches of existing treaties have stood up to close examination. Given the level of interpower rhetoric, the observance of agreements—even those that remain unratified—has been remarkably good.

My confidence that the problems of verification are soluble must be shared by such well-advised people as Ronald Reagan and Margaret Thatcher because they would not otherwise have been sincere in their proposals for the so-called zero option plan to remove all intermediate-range ballistic missiles from Europe. This plan would depend heavily on instant and absolute verification and would lead to catastrophic consequences if cheating were to occur.

In recent months I have discussed the cake-sharing scheme with every disarmament expert and official willing to listen. I have been surprised to discover how many of them expect that the major stumbling block may prove to be getting the armed services of each side to agree internally about military value percentages. My innocent views about the convenience of discussion in the same language, devotion to duty, military discipline, and patriotic loyalty to the supreme commander have been badly shaken. I have found it disturbing to learn that the generals are as wary of the admirals as they are of the enemy, and that this suspicion will intensify if military spending is reduced. We have to devise a scheme for resolving internal wrangles before tackling the larger task.

Clearly, the problem must be soluble: every year we share the military budget among the services and choose to spend money on one weapon rather than another. These choices are declarations of our current perception of the relative values. One rule for writing the lists would be to base military value percentages on the original costs of the weapons, with allowances for inflation and obsolescence. Presidential dictate seems tyrannical.

A more useful method uses a technique that has proved so successful in the thorny disputes between trade unions and employers that it deserves wider recognition. The normal procedure is for the unions to ask for a grossly exaggerated pay increase while the employers respond with a minuscule offer. The parties eventually converge on a settlement between the two initial extremes. Each party gains honors from its constituents in proportion to how close the final settlement is to its starting gambit. The convergence can be assisted by a mediator, but his presence is not essential.

A less common arrangement requires the presence of a neutral arbitrator whose aim must be the benefit of both parties. The arbitrator is allowed to choose one or other of the proposals in its entirety with absolutely no modifications. Because the more outrageous proposals stand less chance of being chosen by the arbitrator, each side competes to make its proposal appear the more reasonable, and so the initial positions tend to become virtually identical. Mediation and arbitration are quite distinct.

Taking arbitration as a model, each armed service would write out its list of military value percentages for the inventories of all the services, supporting them with carefully written advocacy. The supreme commander, taking whatever advice he pleases, would accept one list for the next round of reductions. There would be no need for the public to know which lists had been rejected. The services' anxiety about the decision need not be high because of the very small steps involved. The services that had their lists rejected could modify them for the next round, learning by experience. The lists would be so similar, however, that the selection of a rival one would not need to be perceived as a humiliation.

Other disarmament experts with whom I have spoken prefer a nuclear freeze to cake sharing. There can be no doubt that a total freeze on development, testing, production, and deployment of nuclear weapons would go a long way toward relaxing international tension. Unfortunately, the forces opposed to a freeze are very strong. Jobs, promotions, prestige, and prodigious amounts of money are at stake. These pressures are enough to corrupt the flow of information and manipulate political processes.

Some analysts object to cake sharing by arguing that some types of new weapons should be developed if they provide security at lower levels of threat. The cake-sharing scheme, however, can operate even with the introduction of new weapons. Side A would propose a new weapon and suggest a military value percentage for it. This addition would raise Side A's total to more than 100; thus, Side B would be given a free selection to restore the balance. The value side A assigns to the new weapon would have to be large enough so that Side B would see advantage in choosing to reduce the older stock of Side A's weapons rather than immediately selecting the new weapons. This process would encourage the evolution of weapons having high security values to the owner but low threat values to the potential target.

Finally, some experts raise questions about the role of the secondary nuclear powers. The nuclear inventories of the superpowers are so much larger than those of other nations that the smaller holdings pose no immediate difficulty. As disarmament continues, however, their significance will grow. At some point in the future, when the superpowers have disarmed in a major way, the smaller nuclear powers should participate in the cake-sharing disarmament process.

In sum, cake sharing offers the best hope for overcoming the barriers to arms reduction. It makes reductions through many safe small steps rather than requiring a few risky big ones. It thrives on the superpowers' differences of opinion on the security value and threat value of weapons, thereby reducing the number of high-threat weapons but leaving in place the high security ones. Thus, at each stage of a cake-sharing reduction, both sides can feel they are gaining. While cake sharing will face certain problems—the introduction of new weapons, dealing with small nuclear powers, verification—these can be overcome. The case for cake sharing is so strong that only those who believe their side has nuclear supremacy and are determined to retain it can have logical reservations.

Finding Effective Lower-Risk Means of Defense

We should press the leaders of the superpowers to answer the following crucial question: "Despite your own feelings about the malevolence of your opponents, how big a cake-sharing slice would you risk in order to demonstrate your goodwill?"

1. I am deeply indebted to Richard Garwin for his valuable criticism and encouragement. I cannot claim to be the first person to think of cake-sharing solutions to disarmament problems. The real credit must go to some long-forgotten child whose intelligence allowed him (or more probably, her) to overcome a dangerous aspect of instinctive behavior. The earliest mention that I can trace in the literature is by E. Singer in "A Bargaining Model for Disarmament Negotiations," *Journal of Conflict Resolution* 7 (1963): 21–25. The idea has been developed by F. Calogero. See Calogero, "Some Remarks and a Proposal Concerning the Limitations of Strategic Armaments," *Proceedings of the 22nd Pugwash Conference* (Sept. 1972): 305–17; "A Scenario for Effective SALT Negotiations," *Science and Public Affairs* (June 1973): 17–22; "A Novel Approach to Arms Control Negotiations?" *Proceedings of the 27th Pugwash Conference* (Aug. 1977): 1–10.
2. S. Gorovitz, "When Both Bidders Lose at a Stupid Game," *International Herald Tribune,* Aug. 16, 1983.
3. L. E. Dubins and E. H. Spanier, "How to Cut a Cake Fairly," *American Mathematical Monthly* 68 (1961): 1–17; K. Rebman, "How to Get at Least a Fair Share of the Cake," in *Mathematical Plums,* ed. R. Honsberger (New York: Mathematical Association of America, 1979): 22–37.
4. It is interesting to consider the effects of deliberate distortion of military value percentages. Random deviations on either side of your own best perceived values are of no benefit to you. They give your opponent the chance to obtain an excess benefit by picking a greater number of undervalued good weapons. Instead, you want your opponent to select your overvalued but inferior weapons and leave behind more of your better ones. Distortion should be a legitimate ploy, but it is not without risk. One cannot increase the value of a particular weapon without reducing that of another. If the espionage services of the other side are up to their task and detect the distortion, the gambit will backfire. I believe that the best policy may be to make one's lists an accurate reflection of one's own view.

   An interesting variant of the value percentage scheme has been proposed by Robert Apfel of Yale University. He suggests that Side B should assign military value percentages to the weapons of Side A, who would then choose any items from its own personal arsenal totaling the agreed-upon disarmament step. This means that the actual items selected are chosen by their owner, but that the number of weapons removed is chosen by their potential target. The essential separation of selection and valuation remains. It may be that this variant would make military leaders feel more in control of their remaining weapons. However, political leaders and the civil defense authorities may feel happier about influencing the weapons aimed at their own populations.

Say it were not possible to restrict offensive weaponry in the manner urged by the Delhi Declaration, proposed by Kennan, or imagined by Salter, or there were delays in doing so, or the need to defend territory would remain even after the number of offensive arms were reduced. Say further that computers make it possible to build various small, agile, smart weapons that can defend against big offensive weapons. As Freeman Dyson says in *Weapons and Hope*,* winner of the National Book Critics Circle Award, it would be David against Goliath all over again.

For a while, technology has not favored defense. It produced ballistic missiles but not reliable, cost-effective antiballistic missiles. It produced nuclear-powered missile submarines, but no certain way to keep track of the locations of enemy subs. As a result, military strategy shifted from defending assets in the direction of retaliating against the enemy's assets. What really needed to be "defended" was the ability to strike back, as in the case of missile silos. Instead of expecting to stop an enemy from hurting us, we came to rely more heavily on the threat of hurting him—not only hurting, but probably destroying.

In the next three chapters we look at defense, an old idea whose time may be coming back. As the public support for President Reagan's Strategic Defense Initiative reveals, people dream of having a defense again, as America was once defended by the oceans. Many Americans resonate to Reagan's image of erecting an invisible shield above the U.S., off which, in our imagination, missiles would bounce.

In Dyson's version of missile defense, small rockets would tear apart enemy missiles not during the "boost phase," over their silos, but during the descent just minutes before they would otherwise explode. In Dyson's "fantasy," the tiny defender would work simply by smashing into the big incoming missile and destroying it. The whole trick would be finding its way to exactly the right place. That's the job of the computers.

If it were possible, this defense would certainly change the nature of deterrence, but it faces the same problem as Reagan's Star Wars ambition—even a single missile that got through would do horrendous damage. In the case of defending against a tank attack, however, a miss here and there by "precision guided munitions" would not cost a city.

Why not arrange things in Europe, Dyson says, so that each side could stop the other in its tracks on the ground? In a second stage the sides would reduce the number of tanks they kept, further diminishing

*Freeman Dyson, *Weapons and Hope* (A Cornelia & Michael Bessie Book/Harper & Row, 1984).

the possibility of a successful attack by either. Disturbed by NATO's current reliance upon "tactical" nuclear weapons to blast away at a Soviet breakthrough, Dyson calls for "weapons that are capable of defending territory without destroying it in the process." You'd think the Germans would be interested.

David and Goliath

Freeman Dyson

And there went out a champion out of the camp of the Philistines, named Goliath, of Gath, whose height was six cubits and a span. And he had an helmet of brass upon his head, and he was armed with a coat of mail; and the weight of the coat was five thousand shekels of brass. . . .

And the Philistine said to David, Come to me, and I will give thy flesh unto the fowls of the air, and to the beasts of the field. . . .

And David put his hand in his bag, and took thence a stone, and slang it, and smote the Philistine in his forehead, that the stone sunk into his forehead; and he fell upon his face to the earth.

THE STORY OF DAVID WITH HIS SLINGSHOT SLAYING THE clumsy giant has delighted children for three thousand years. David was an early example of a common type of folk hero, the boy who fights with skill and daring against superior force and wins. Some time before David's triumph in the valley of Elah, Odysseus was in Sicily winning his battle of wits against the Cyclops. Similar stories are found in the folklore of nations all over the world.

A modern version of the same story might run as follows: One fine day, out of a blue sky, a monster ballistic missile carrying a megaton warhead comes falling toward its target at a speed of three miles per second. Five minutes before the missile is due to arrive, a little rocket weighing a few pounds goes up from the ground to meet it. The little rocket carries no warhead and does not move very fast. It carries only a telescope and a microcomputer and an accurate servosystem to bring it onto the missile's track. The missile crashes into the rocket at three miles a second, at which speed the rocket tears through the warhead and rips it apart. The missile is destroyed by its own speed; the rocket has to do nothing except to be there at the right place at the right time. The dead missile falls upon its face to the earth.

This little story might come true, at some experimental missile-testing range, with the megaton warhead replaced by a suitable dummy. It might happen at an American range, or at a Soviet range, or both. Let us suppose for the sake of the story that it happens at an American range. Then the experiment does not remain secret for long. Films of the encounter appear on television. There are articles in *Aviation Week* describing the event and making exaggerated claims for the effectiveness of the interceptor. There are articles in more sober magazines, pointing out that a successful demonstration of an interception of a single missile over a test range does not prove the feasibility of non-nuclear defense against a real attack by hundreds or thousands of missiles. Experts testify before congressional committees that the development of non-nuclear interceptors is a waste of money, that the staging of the test interception is a propaganda exercise calculated to deceive the public. But the public is not deceived. The public understands well enough the difference between a test interception and an operational defense. The public also understands, better perhaps than the technical experts, that the test interception is an event of some importance.

A successful David-and-Goliath experiment, with a two-pound rocket killing an intercontinental missile, even under the artificial conditions of a test range, could be a historical turning point. It could mark the beginning of a change in the way people think about nuclear weapons. It would show nuclear weapons in a new light, as clumsy brutish things outwitted by a cheaper and more agile adversary. The death of Goliath could be the death of an era, the end of the unquestioned technical supremacy of nuclear weapons. It would make clear to everybody that we have a choice, either to continue relying on the old technology of nuclear weapons, or to switch our emphasis to the newer and more versatile technology of sensors and computers. The victory of the two-pound interceptor could be a signal to the world that nuclear weapons are at last on their way down, that the complete abolition of nuclear weapons is technically thinkable.

The point of the David-and-Goliath story is not that David will always win. On the contrary, it is easy to imagine Goliath winning if he uses suitable countermeasures against David's tactics—for example, by redesigning his helmet of brass so that it covers his vulnerable forehead more effectively. The point of the story is that David won once, and that was enough to make Goliath a laughingstock for three thousand years. After Goliath's defeat, giants no longer commanded respect. After the technological supremacy of nuclear weapons has once been challenged, their moral and political unattractiveness will stand revealed clearly to the eyes of the world.

Another modern version of the David legend would put a bomber airplane or a cruise missile in the role of Goliath. David would again be a small agile rocket with the ability to intercept and kill by direct impact. David's task in this case would be technically more demanding, because bombers and cruise missiles do not fly straight. Again, David will not always win, but if he is small enough and cheap enough, he can make his enemies look like blundering fools.

A third variation on the David-and-Goliath theme could be staged in the ocean. In this version of the story, Goliath is a billion-dollar nuclear submarine carrying long-range missiles with enough warheads to destroy a whole country. David is a little mechanical suckerfish, programmed to attach itself to the skin of a submarine, imitating the remora suckerfish, which attach themselves by suction to the skins of sharks. David lurks in the water until Goliath approaches, then swoops down and quietly sucks. Once attached to Goliath, David computes his exact location and is ready to transmit this information to headquarters, either at predetermined times or upon demand. The trick of using suckerfish to catch submarines is not a modern invention. It has been known to the Australian aborigines of Cape York since time immemorial. The aborigines used the trick with captive suckerfish (gapu) to catch sea turtles. "The same gapu will catch three or four turtle," reports a nineteenth-century observer. "When they have brought it up, they take the gapu off and put it out to catch another."

This story of the suckerfish raises troublesome questions. Both the United States and the Soviet Union have invested enormous resources in their fleets of missile-carrying submarines. The strategic doctrine of the United States relies heavily on the ability of submarines to remain undetected. According to the orthodox doctrine of deterrence, the undetectability of submarines provides our strongest guarantee of strategic stability; so long as we have a major part of our offensive force securely concealed in the ocean, nobody can hope to gain any decisive advantage from an attack on our more vulnerable land-based forces. Conversely, if submarines become detectable, the existence of stable deterrence is endangered and the peace of the world becomes precarious. This is the orthodox doctrine which has guided United States policies for the last twenty years. Many of our experts in arms control believe so firmly in this doctrine that they look to the missile-carrying submarine as a uniquely benign form of strategic armament. Whether or not we agree with the official doctrine, we must recognize that the missile submarine has many technical and political advantages. The loss of submarine invulnerability would force us to make radical changes in our strategic assumptions. In short, the story of the suckerfish may be not merely troublesome but dangerous. Perhaps it is irresponsible of me to mention it as a possibility.

What can I say in defense of the suckerfish? I will say three things. First, the story is only a story. So far as I know, no such animal exists, even as a diagram on the desk of an inventor. I tell the story only as an example of the consequences which the technology of microcomputers may bring to military operations in the ocean as well as on land. Second, I am not suggesting that anybody should attempt to attach suckerfish to operational Soviet submarines; this would indeed be a highly provocative and dangerous thing to do. Third, it would take enormous numbers of suckerfish distributed over huge areas of ocean to ensure that all missile-carrying submarines were located; such a massive deployment of mechanical fish could not be carried out quickly, cheaply, or secretly; the technology of suckerfish is not about to take the world by surprise. But after these disclaimers have been made, it remains true that the technology of suckerfish, or some similar non-nuclear technology, could in the long run put an end to the popularity of missile-carrying submarines. For David to defeat the submarines, it would not be necessary for him to locate all the submarines all the time. He needs only to throw serious doubt upon the ability of submarines to remain concealed. If submarines cannot rely on concealment, the military advantage of using them as missile carriers disappears. A successful demonstration of suckerfish technology would reveal the missile-carrying submarine for what it is, a vulnerable Goliath carrying far too many lethal eggs in one fragile basket.

When mankind decides, as I believe we must ultimately decide, that missile-carrying submarines are intolerably dangerous, the task of eliminating them will be carried through sooner or later, with or without the help of suckerfish. The task will be made easier if the military virtues of submarines have already been discredited. A robust technology of submarine location will allow an international agreement banning sea-based missiles to be effectively monitored. As always, the primary requirement for carrying through any act of nuclear disarmament is the political will to do so, but the formation of such a will can be powerfully helped by a technological development deliberately aimed toward making nuclear weapons unattractive.

Let me now return from fantasies to facts. There exists a vigorous and rapidly advancing area of military technology known as precision guided munitions or PGM. PGM are small accurate missiles with nonnuclear warheads, mainly designed to kill tanks or airplanes. They are of many kinds, some small enough to be fired by infantry soldiers, others carried in armored cars, helicopters, airplanes, and ships. They were first seen in action on a large scale in the 1973 Middle East war, when PGM supplied by the Soviet Union to Egypt and Syria caused heavy losses to Israeli tank forces. Since 1973 the technology of PGM has been pushed much further, not only in the Soviet Union and in the United States but in other countries too. Experts disagree concerning the extent to which PGM have revolutionized land warfare. Some say that PGM will defeat tanks in the 1980s as decisively as airplanes defeated battleships in World War II. Others say that tanks will still survive provided that they are supported by friendly PGM and infantry. Nobody doubts that PGM have put an end to the era in which tank armies won easy victories.

The existing PGM technology fits well into the David-and-Goliath tradition. The PGM themselves are light and agile and comparatively cheap; the tanks which they destroy are heavy and clumsy and expensive. The PGM exploit the new technology of accurate sensors and servos and computers; the tanks are stuck with the old technology of armor and guns. The PGM are likely to prevail in the end because they are more cost-effective. But there is always a chance that PGM will fail, especially if the people who design them succumb to the temptation to improve them until they become as complicated and expensive as tanks.

The emergence of PGM technology has raised questions of fundamental importance for the feasibility of nuclear disarmament. Can PGM be an effective substitute for tactical nuclear weapons? Is it possible to use PGM as the basis for a credible non-nuclear defense of Western Europe? In answering these questions, it must be said at the outset that the success of PGM in stopping a Soviet tank army from rolling across Germany cannot be guaranteed. Effective use of PGM requires large numbers of competent and brave soldiers. If equipment alone could guarantee a successful defense, France would not have been overrun by a numerically inferior force of German tanks in 1940. Warfare is always a gamble, for the defenders as well as for the attackers. It makes no sense to demand that PGM provide a defense with zero risk of failure. The question we should be asking is whether the risks of a PGM defense would be smaller than the risks of our present reliance on tactical nuclear weapons. This question cannot yet be answered with certainty, but there are strong reasons to believe the answer will be affirmative.

The chief virtue of tactical nuclear weapons is that they make a European war unthinkable and allow the Europeans to avoid worrying seriously about defense. So far as Germany is concerned, the tactical weapons are almost the equivalent of a Doomsday Machine; in case of any major trouble, Germany is destroyed almost automatically. Like a Doomsday Machine, the nuclear weapons bring stability and tranquility to the political life of Europe, and save us a great deal of money which would otherwise have to be spent on non-nuclear armaments. Sudden withdrawal of the nuclear weapons would be politically upsetting and militarily risky. Nevertheless, the argument which Herman Kahn used against his Doomsday Machine applies with equal force against tactical nuclear weapons; they are not acceptable as a permanent solution of the problem of European security because a failure of the system kills too many people and kills them too automatically. In the long run, the risks of a PGM defense are smaller. The problem is how to manage the transition from nuclear to non-nuclear defense without running into worse risks in the intermediate stages.

It is not likely that the denuclearization of Europe can be achieved by substituting PGM for tactical nuclear weapons directly. The substitution will probably come about indirectly, through a sequence of preliminary steps. One preliminary step might be a partial replacement of tanks by PGM in the Warsaw Pact forces as the Soviet authorities become aware that tanks are no longer cost-effective. Another intermediate step might be a negotiated trade-off of Soviet tanks against NATO nuclear weapons. A third step might be an agreement to eliminate the remaining tactical nuclear weapons on both sides. It is easy to invent other possible sequences of steps leading to nuclear disarmament. The essential point is that the decisive last step, the agreement to get rid of the nuclear weapons, will be made much easier by the preliminary downgrading of tank forces in response to PGM development.

If the elimination of tactical nuclear weapons proceeds in this indirect fashion, we are not at any stage left in a situation where the existing Soviet tank armies are in Central Europe and only a few NATO soldiers with PGM stand between them and the Rhine. Instead, we have a situation in which tank armies on both sides diminish as PGM become more capable. Our aim must be to achieve simultaneously the withdrawal of nuclear weapons and the obsolescence of tank armies. It is important to understand that this program cannot succeed if only the Western armies deploy PGM. A heavy Soviet investment in PGM is also essential. In other words, if we want to free both East and West from addiction to tactical nuclear weapons, the best way may be to engage both sides in a PGM arms race.

The mathematician John von Neumann played a leading part in the development of nuclear weapons and an even larger part in the development of computers. He said in 1946, when both these developments were just beginning, that the computers would be more important than the bombs. The impact which computers have made upon civilian society has already proved him right. And now the advent of PGM promises to prove him right in the military sphere also.

The computer revolution transforms war into a contest of information rather than of brute force. It enables small cheap devices with brains to overwhelm big expensive vehicles. It favors David against Goliath. At sea, aircraft carriers become vulnerable to small boats carrying torpedoes or cruise missiles. On land and in the air, tanks and airplanes become vulnerable to PGM. In space, it is no longer absurd to imagine smart little non-nuclear rockets killing nuclear missiles. In all four places, the computer revolution is chipping away at the supremacy of nuclear weapons, providing the technological foundations for a world in which weapons of human scale and purpose take the place of weapons of mass destruction.

Another hopeful aspect of the computer revolution is that it tends in the long run to shift the balance of forces away from offensive and toward defensive weapons. The qualifying phrase "in the long run" is here essential; two of the first fruits of the computer revolution were the cruise missile and the MIRV, two weapons which further strengthened the supremacy of the offensive. But in the long run, as war becomes increasingly a contest of information rather than of firepower, defense becomes easier and attack more difficult. Already today the effects of PGM in land warfare are favorable to defense. In land warfare, offensive units must be mobile and visible, hence vulnerable to PGM, while defensive units can be concealed and relatively invulnerable. The fundamental reason why the computer revolution favors defense is that in a battle of information, the defenders fighting in their own territory can see what is happening better than the attackers fighting in exposed vehicles. Even in the strategic contest of missile against antimissile, the same principle should be valid in the long run. If the antimissiles are numerous enough and agile enough, they can destroy any missile which they can clearly see, and the defenders directing the antimissiles over their own territory will have a clearer view of the battle than the attackers sitting 5,000 miles away.

Finding Effective Lower-Risk Means of Defense

The slow shift of technological advantage toward defense does not imply that a country will be able to survive a nuclear attack without catastrophic damage. It implies only that there will be some economic and strategic incentives for transferring resources from offensive to defensive deployments. Defensive technology cannot by itself ensure our survival. But it can give useful support to the political movements and arms control treaties which will finally put an end to the technology of mass destruction.

All through human history, military technology has appeared in two guises, with the face of Goliath and with the face of David. The contrast between the two faces of weaponry became sharper than ever with the advent of modern technology. Our century has provided spectacular demonstrations of Goliath and David in action. Many observers of the human situation, seeing what our technological Goliaths achieved in the two world wars and later in Vietnam, have come to believe that all modern weapons and military establishments are instruments of enslavement, natural enemies of the independent rational soul, and that a morally responsible person should have nothing whatever to do with them. Einstein expressed this view eloquently in his writings and in his actions. Yet one cannot begin to understand the deep involvement of American scientists in military technology if one does not examine the contrary view, that freedom and military inventiveness are natural allies.

John von Neumann was the most brilliant and the most articulate of the scientists who consciously devoted their talents to the improvement of weaponry in the cause of freedom. Von Neumann's generation saw free societies obliterated all over Europe, not by internal forces of oppression but by Hitler's armies. Freedom survived in England in 1940 because the technological Davids, the coastal radars and the fighter airplanes, were there when they were needed. Many people at that time believed that freedom's survival was made possible by the willingness of British and American scientists to apply their skills wholeheartedly to the problems of war. Even Einstein shared this belief in 1939 when he helped his friend Leo Szilard to launch the American nuclear energy enterprise. After Hiroshima, Einstein changed his mind, but the majority of American scientists did not.

Their experience of World War II left behind it a widespread feeling that a permanent alliance between freedom and military science was right and proper. The alliance was evidently beneficial to both parties; a free society needed superior military technology to withstand the superior discipline of a totalitarian enemy, and the military establishment needed a free society to allow scientists and soldiers to work together in an informal and creative style which a totalitarian state could not match. In the context of the Soviet-American arms race, the free scientists of America carried a responsibility to stay ahead in the quality and variety of their inventions, so as to compensate for the larger military expenditures and the advantages of secrecy on the Soviet side. This picture of the arms race sounds naive and old-fashioned today, but it was dominant in von Neumann's thinking; it still flourishes in Israel and in the less sophisticated regions of America. For every scientist who believes with Einstein that modern weapons in the hands of modern governments are an absolute evil, there is another who believes with von Neumann that modern weapons rightly used can help David to survive in freedom in a world of Goliaths.

The root of the moral dilemma of our age lies in the fact that both Einstein's and von Neumann's viewpoints contain a large element of truth. Einstein was right in saying, "Mankind can gain protection against the danger of unimaginable destruction and wanton annihilation only if a supranational organization has alone the authority to produce or possess these weapons." Von Neumann was right in believing that the old political realities of national power and the old tribal imperatives of fighting for survival would remain essentially unchanged even in a world armed with hydrogen bombs. So long as nation-states continue to base their security on threats of general annihilation, there can be no escape from the moral dilemma. But a way of escape may lie open if nation-states can find military means which are effective both in defending tribal interests and in diminishing reliance on nuclear weapons. Each nation is exposed to two kinds of danger, the danger from particular international conflicts and the danger from nuclear weapons in general. An optimum strategy should be designed to deal with both dangers simultaneously. We should recognize explicitly in our military doctrine that our own reliance on nuclear weapons creates a threat to our security which is as serious as any external threat. This being accepted, the appropriate weapons to implement our military doctrine will be the weapons of David, weapons which are capable of defending territory without destroying it in the process. The way out of the moral dilemma is to demand that our weapons answer mankind's need for deliverance from nuclear threats as well as our tribal need for defense of our friends' territory.

All our military arrangements should be designed with two purposes in mind: the short-range purpose of discouraging aggressive countries from overrunning their neighbors, and the long-range purpose of achieving a gradual transition to a non-nuclear world. The two purposes are not incompatible. The weapons which are most effective in wars of local territorial defense are likely also to be helpful in the global political war against nuclear annihilation.

Einstein believed in no such compromises. He said, "I would unconditionally refuse all war service, direct or indirect, and would seek to persuade my friends to adopt the same position, regardless of how I might feel about the causes of any particular war." On another occasion he said, "I appeal to my fellow scientists to refuse to co-operate in research for war purposes." Perhaps he was right. Perhaps the hope of using non-nuclear technology to make nuclear weapons obsolete is an illusion. But the example of David keeps the hope alive.

Nobody can predict the future of technology over a long time scale. But it may be possible to make predictions which will be valid for a decade or two. It seems likely that the rapid development of microcomputer and sensor technology will result in a growing proliferation of sophisticated non-nuclear weapons of the kind which I described in my David stories at the beginning of this chapter. The dissemination of David weapons will cause armies to take a step back into an older, more professional style of warfare. The new weapons need elite, highly trained soldiers to use them effectively. They do not need the mass armies which provided the cannon fodder of the two world wars. The Falklands campaign of 1982 provides some additional evidence that the winds of change are blowing in this direction. The Argentine air force, a small elite force using precise weapons with daring and skill, did great damage to the invading forces, while the

Argentine army, a mass army of conscripts, was crushingly defeated. It seems that modern technology is taking us back toward the eighteenth century, toward an era when small professional armies fought small professional wars.

How lucky it would be for humanity if the new David weapons were to make obsolete both of the two great monstrosities of the twentieth century: the nuclear mass-destruction weapons and the great cannon fodder armies! That would be an unbelievable stroke of luck, as if David were to slay two Goliaths with one stone. History teaches us not to expect such luck, and common sense teaches us not to rely on it. Nevertheless, the obsolescence and ultimate abolition of nuclear weapons and mass armies is a worthy political goal which we should strive to reach by all available means. The technology of David weapons is only one of the available means, and probably not the most important, for reaching this goal. If the David weapons can help even slightly to make the goal politically attainable, they will have served a useful purpose. What the world needs in order to be saved is not technological magic but a rebirth of hope.

As a member of a study group called the Exploratory Project on the Conditions of Peace (EXPRO), Mark Sommer has written *Beyond the Bomb, which is described as "a field guide to alternative strategies for building a stable peace" and from which the following paper is drawn. When he's not traveling to interview defense experts, confer with EXPRO colleagues, or give a speech, Sommer lives with his wife on a homestead in rural California, connected to the modern world only by a "bush telephone."**

Under these conditions, he is able to elude the assumption that it's reasonable for nations to threaten one another, perpetually, with nuclear destruction. To Sommer, apple trees are reasonable. So are the moon and stillness. After reading somebody's article about "nuclear war-fighting," it's a relief to turn to an essay by Sommer and feel sanity blow in like a breeze off the hills. He never confuses security with an ability to blow up somebody else's place.

EXPRO currently consists of twenty-four men and women, nearly all Americans, either in academic life, policy institutes, foundations, or political action groups, who "believe that the ultimate necessity is to replace the war system now dominating our lives with a workable peace system." Without intending it, we selected four of its members as contributors to this book and its companion volume, *Citizen Summitry.* Apart from Sommer, they are Elise Boulding, Daniel Deudney, and Patricia Mische. With inspiration from W.H. Ferry and under the chairmanship of George Rathjens, EXPRO is quietly and effectively seeking "to promote and identify the elements" of a peace system.

Alternative Security Systems

Mark Sommer

Offense sells tickets.
Defense wins games.
—Motto of the Pacific Northwest Basketball Players Association

OFFENSE AND DEFENSE HAVE PLAYED OUT A CURIOUS cat-and-mouse relationship in the history of warfare. While the attack has always held sway in the human imagination, there have also

***Mark Sommer (for the Exploratory Project on the Conditions of Peace). *Beyond the Bomb: Living Without Nuclear Weapons* (EXPRO/The Talman Company, 1985).**

been stunning episodes of heroic defense—the Battle of Britain, the Warsaw Ghetto uprising, the siege of Leningrad, to name only the most recent. In both technology and tactics, offense and defense have tracked one another in a desperate pursuit of elusive unilateral advantage, the offense probing for a hidden weakness, the defense seeking hidden strengths. In various eras and circumstances new weapons and new strategies have temporarily thrown the balance from one to the other. But at no time has the offense so overwhelmed the defense as since the advent of nuclear weapons.

Ironically, it is also nuclear weapons, those paramount instruments of offense, which have taught us both the insufficiency of defense and the absurdity of attack. Yet both superpowers have continued to act as if a more devastating offense were in fact the only defense. The term "defense" has thus been so muddled by misuse that most Americans—and probably Soviets as well—can no longer imagine a defense that does not rely almost solely upon apocalyptic threats. Yet one need not be a prophet to recognize that there is no long-term stability to be found in a system of mutually induced terror. As threats echo and escalate, we may find that we have not *deterred* war but merely *deferred* it until the results are assured to be utterly cataclysmic.

Recognizing the peril and futility of a system of ever-enlarging threats, a number of theorists, first in Western Europe and more recently in the United States, have begun to ask themselves whether and how a defense might be fashioned that does not depend on threats and yet does not threaten one's own security. The past several years have seen the emergence of a new body of theory lying somewhere between disarmament and arms control. Many of those who now design alternative military strategies have migrated from their first and more familiar homes in the nonviolent tradition. A number of the theorists who first explored the potentialities of civilian resistance have, a decade later, chosen to focus on more intermediate possibilities. In most cases the shift has been guided by pragmatic considerations: civilian resistance was thought to be a "better idea" but simply beyond reach, at least within the next generation, and then perhaps only in those few nations outside the strategic equation of the superpowers, among a highly evolved and homogeneous people.

Finding Effective Lower-Risk Means of Defense

In choosing to examine alternative forms of military defense, these researchers have mostly tried to apply the principles and spirit of nonviolence to the hardware and strategy of armed force. All are aware of the compromises this course entails, the extent to which nearly any use of existing military hardware is destructive and violent in nature. Granting this limited set of choices, they have sought to draw a distinction between offensive and defensive capabilities, between destructive and protective strategies. They are endeavoring to reclaim the original meaning of the term defense from the inverted definition it has been given by post-war nuclear strategists: defense not as the capacity to inflict attack but as the capacity to repel it. These new strategies for a defense without threats have gone by a variety of names—"defensive deterrence," "defense without offense," and "non-provocative defense," to name but a few—but the principles underlying them evince a remarkable similarity. They include:

1. Greatly reducing our reliance on nuclear weapons, both strategic and tactical;

2. Greatly increasing our reliance on weapons more useful for the defense than the offense, making ample use of recent advances in defensive weapons technologies like precision-guided munitions;

3. Developing tactical military strategies strong in the defensive mode and deliberately weak in the offensive, to signal adversaries that they need not fear attack, yet that they will not succeed if they themselves attack. A no-first-use declaration by NATO, accompanied by a shift from tactical nuclear weapons to short-range precision-guided munitions, is one such strategy that has been given much consideration;

4. Adding a variety of non-military and nonviolent forms of defense, from prepared civilian resistance to economic sanctions and incentives and diplomatic initiatives;

5. Confining the task of defense to the territory, people, and institutions of one's own country and eschewing most claims beyond one's borders.

Most of this new thinking has emerged from Western Europe—from Scandinavia, The Netherlands, West Germany, and Great Britain—where it has been seen in part as a means of supplanting the continent's hazardous dependence on nuclear weapons to deter a putative Soviet attack. Mainstream Western strategists have also been considering decreasing their reliance on tactical nuclear weapons for the defense of Western Europe, most notably in the ESECS study, *Strengthening Conventional Deterrence in Europe,*[1] but the thrust of their strategy is quite different. Instead of the strictly defensive posture advocated by alternative defense theorists, NATO Commander Bernard Rogers and others propose "deep strike" strategies utilizing advanced conventional weaponry to interdict Warsaw Pact second echelon troops before they reach the front. The fact that precision-guided munitions play a role in this conventional offensive demonstrates the troublesome convertibility of weapons. Proposed as a technological centerpiece for most alternative defense strategies, the anti-tank weapon can in fact be employed in the service of the attack, if the will is there to do it. The capacity for offense has not yet been bred out of them.

The most comprehensive and ambitious effort to translate the principles of a non-threatening defense to the particularities of a single nation has been the work of the British Alternative Defence Commission. An independent committee of defense and disarmament specialists, with a smattering of churchmen, MP's, trade unionists and others, the Commission spent two years examining the question of how Britain could defend herself without nuclear weapons—both her own and those of the United States. Their conclusions, set forth in a volume entitled *Defence Without the Bomb,*[2] call for a blending of a territorial military defense and a "fallback" capacity for civilian resistance. The Commission is currently embarked on a second two-year study of a foreign policy to match its reformed defense posture.

It is a measure of the success of the Commission and other groups in seeding their ideas in the political mainstream that Britain's Labour Party recently adopted a non-nuclear stance as its official defense policy. The plan, drafted by a committee of both left- and right-wing members of the party, proposes that Britain scrap its Trident submarine program, decommission its Polaris missile system, and remove American military bases from British soil. While Britain would remain in NATO, it would "ask the United States and NATO as a whole to hold negotiations with Britain about the latter's withdrawal from the West's nuclear defense planning arrangements. The wording of the policy document is "uncompromising," reports *The Christian Science Monitor.* "Membership of (sic) NATO does not require member states to have nuclear weapons of their own or have U.S. nuclear weapons on their territory, or to accede to a strategy based on nuclear weapons. (Neil) Kinnock [new leader of the Labour Party] has pledged himself to support a policy line that would give Britain a leading role in advocating establishment of a non-nuclear defense for NATO . . . . Already it is obvious that if Labour comes to power, a serious crisis on defense matters with the United States is probable."[3]

Frank Barnaby and Egbert Boeker have devised in general terms a strategy of non-nuclear defense for the entire NATO Alliance. They propose the use of advanced conventional defensive weaponry in place of nuclear weapons (precision guided munitions, electronic sensors, remotely piloted vehicles, etcetera). Indeed, their faith in the efficacy of a technical defense exceeds that of many of their colleagues. They also propose a gradual phasing-in of an East-West treaty of non-provocative defense, with intermediate steps to proscribe first use, and then to forbid possession of nuclear weapons. This suggestion bears a resemblance to proposals for "qualitative disarmament" first conceived between the two world wars. "European security is possible without nuclear weapons," Barnaby and Boeker conclude. "It is impossible with them."[4] Barnaby and several retired NATO officers have established an organization aptly titled, "Just Defence" to press for adoption by NATO of a non-provocative defense posture.

The most comprehensive and detailed study to date of an alternative defense posture for the United States is the report of the Boston Study Group, a small colloquy of scientists and defense and disarmament specialists (including Randall Forsberg, author of the Freeze, and Philip Morrison, a physicist and participant in the Manhattan Project) which undertook a five-year examination of the American military budget and arsenal. First released in 1979 as *The Price of Defense* (and republished in 1982 as *Winding Down: The Price of Defense*), the report calculated that the United States could safely subtract 40 percent from its defense budget without endangering its security. "The alternative military policy which we propose," the report states, "is not only safe, but actually safer than the present policy, even if there is no corresponding, constructive change in other countries."[5] The group specifies three categories it seeks to *reduce:*

1. Excess nuclear weapons in the American arsenal, including all weaponry useful for a first strike;

2. Most of the aircraft carriers, amphibious landing craft and lightly equipped land combat forces most useful for interventions in the Third World (although it did not yet exist when the report was written, the Rapid Deployment Force established under President Carter is just such a mobile invasion force); and

3. Investment in the development of new offensive weaponry, both nuclear and conventional.

The forces the Study Group chooses to *retain* include:

1. A small but invulnerable nuclear weapon force sufficient to retaliate but clearly inadequate to threaten a first strike;

2. The heavily equipped land combat forces and tactical combat aircraft to be used in the defense of Western Europe; and

3. A largely unchanged force of surface ships and attack submarines, to protect the freedom of the seas.

More recently, Frank von Hippel and Hal Feiveson, both of Princeton, have initiated a study of the technical issues relevant to a negotiated regime of minimal deterrence. A team of Soviet scientists has undertaken a parallel study under the direction of Academician Evgeny Velikhov of the Soviet Academy of Sciences. The Princeton study is to examine "an alternative future in which nuclear weapons play as little a role as possible in international politics and military planning. Nuclear weapons would still exist, but both superpowers would have adopted no-first-use policies and would retain in their nuclear arsenals only enough weapons to deter their use by other states."[6]

Neither of these studies advocates a defense without offense in quite the sense in which it is being proposed in Western Europe. Both continue to rely on a residual deterrent of mostly submarine-based nuclear weapons, and the Boston Study Group leaves in place a variety of existing alliance commitments. The difference between the approaches simply reflects the greater reach of currently perceived American interests and power and the greater distance that must be traversed in order to reach a genuine defense without threats. American theorists are well aware of how much more difficult a proposition it is to conceive a United States without the Bomb than a Great Britain, for only Britain depends on the arsenal she has built (and then sits beneath the American umbrella for good measure), while entire alliances depend—or believe they depend—on the American nuclear arsenal.

This emphasis on a defense-only military posture has recently been echoed in the proposals of three other well-known theorists. Randall Forsberg has conceived a mode of disarmament in which nations progressively shed the most offensively oriented components in their arsenals and strategies and leave intact only those elements useful for the defense of one's own territory.[7] Jonathan Schell, formulating what he calls a "deliberate policy" to abolish nuclear weapons, chooses to leave in place a conventional arsenal with a strictly defensive orientation.[8] And Freeman Dyson, wearing twin hats as both weapons designer and peacemaker, has urged in two widely read books published during the past several years[9] that the United States, and as many other nations as it can persuade to follow, adopt a "defense-dominated future." "A good military technology," he writes, "is one that leads away from weapons of mass destruction and toward weapons that allow people to defend their homeland against invasion without destroying it."[10] Terming his policy "live-and-let-live," Dyson argues that: "the fundamental change in objective is that we look to a defense-dominated balance of non-nuclear forces rather than to an offense-dominated balance of nuclear terror as the ultimate basis of our security."[11]

Although he urges that the transformation be negotiated, Dyson also allows for the possibility of generating a spontaneous process of benign imitation:

"Our technology, if we care to use it for this purpose, gives us a uniquely effective means for guiding Soviet policies in directions which we may consider desirable. This channel of communication has the advantage of being always open. Soviet leaders do not always wish to listen to our diplomacy, but they always listen to our technology. We cannot use technology to persuade them to move in directions which they consider contrary to their interests. But our technology can influence them effectively whenever, as often happens, we wish them to move in a direction where their interests run parallel to ours. The move to non-nuclear defense is a case in point. If our technologists lead strongly into non-nuclear defensive weaponry, it is a good bet that theirs will follow suit."[12]

Swiss-born peace researcher Dietrich Fischer has gone furthest in conceiving a comprehensive general theory for a defense without threats. Drawing on the example of Switzerland's territorial defense, Fischer advocates "transarmament,"[13] which he defines as a shift from offensive armed force to defensive military and non-military strategies. He terms this shift a "robust solution," "an approach to the prevention of war that works regardless of the actual intentions and capabilities of the opponent, whether he has aggressive aims or seeks cooperation, whether he acts independently from what we do or reacts to our policies, whether his offensive and/or defensive capabilities are strong or weak."[14]

Fischer is especially insightful in his analysis of non-military defense (not to be confused with civilian-based defense) and its relation to a defensive military posture. He specifies four categories of policy that can be adopted:

1. increasing the losses of an opponent if he attacks (through a loss of trade, exchange, prestige, and other benefits he enjoys in peacetime);
2. reducing the gains of an opponent in case he attacks (publicizing in advance one's plans to demolish anything of tactical value to an invading army);
3. reducing the losses of an opponent in case he does not attack (refraining from hostile acts, reducing and/or eliminating threats); and
4. increasing the gains of an opponent in the event he does not attack (through exchanges and trade of all sorts).

Taken together these policies constitute what Fischer calls "dissuasion," "a form of non-threatening deterrence which, without evoking fear, aims at convincing an opponent that keeping peace is better for him than resorting to war."[15]

These theories display a remarkable convergence of perspective in a research community that is not inclined towards consensus. The merits of a purely defensive strategy are manifest. Posing no threat to adversaries, it thus eliminates any provocation or justification for attack. The relinquishment of threats evaporates at least half the fuel for the arms race between East and West. Further, a defensive strategy would greatly reduce reliance on nuclear weapons, which are by nature offensive systems; whether it could eliminate them entirely is an open question. A defensive strategy is ethically more satisfying than MAD in that it returns defense to its original meaning and function—protection instead of destruction. And finally, "defensiveness" does not ignore the possibilities of harm and the requirement of shelter in an often hazardous environment.

But defensiveness is not without its deficiencies, limitations that the theorists themselves may know most acutely. It is no doubt wiser to face these difficulties directly at the outset than to pretend to an infallibility that cannot be promised to any single strategy, any means of defense. Defensiveness taken by itself is an incomplete policy and concept, a reactive rather than a creative strategy. It is too passive, too apparently vulnerable a stance to attract and utilize the expansive energies of nations and peoples accustomed to taking leading roles in world events. We may deplore these assertive energies and the destructive means by which they often vent themselves, but we cannot deny an outlet for the energies themselves. In addition, purely defensive strategies do not meet the challenge at its source but at its outcome, where the full force of the attack has been marshalled and concentrated, placing the defender at a distinct disadvantage, both psychological and strategic. Having stripped itself of all means to press the initiative, a purely defensive strategy appears to lay itself open to mischief.

In foreclosing use of offensive weapons and strategies, a purely defensive strategy may be placing itself at a critical disadvantage in relation to an adversary who does not forswear their use. The abandonment of threats, while a most welcome act in itself, will not wholly eliminate the possibility of threats from the other side. We must be tough-minded enough to anticipate an adversary who will push even when not shoved. This is not worst-case thinking but simple prudence. So a defense without threats must be doubly or triply strong in its purely protective capability, for it must meet an adversary who is undistracted by the need to fend off the blows of another and fully concentrated on the attack. No defensive strategy can promise invulnerability—a recognition that has led even those military planners who seek no territorial claims beyond their own borders to argue the necessity of some means of carrying the battle back to their adversaries.

And finally, there is no simple distinction to be drawn between offensive and defensive weaponry. It might well be observed that one man's defense is another man's threat. The function of a weapon depends as much on how it is deployed as on what it can do. Thus even short-range aircraft become potentially offensive weapons when placed in close proximity to the borders of conflict. Yet they may be designed and deployed for use as interceptors. There is little in the plane itself that prevents its offensive use. Indeed, aircraft designs increasingly include both offensive and defensive functions in order to eliminate the costs of making two different planes. Thus we return to the gauzy realm of intentions, uncertain and incalculable, and to the question of capability versus will. If your adversary has the means to kill you, does it mean he wants to? Most significantly, there is the irony of deterrence itself, an overwhelming threat ostensibly used as a "defensive" measure.

The distinctions between offense and defense will never be as clean as we would wish them to be, but that is no excuse for not making the distinction at all. Though the gray region is large, so is the black. The great majority of nuclear weapons systems clustered under the umbrella of deterrence are in fact instruments of coercion masquerading as those of protection. So although the distinction is not absolute, it is fundamental.

These evident limitations of defensive strategies are not immutable. But to get past them we may have either to expand our notion of defense or to enlarge our repertoire of defensive strategies—or more likely both. In rightly forswearing aggressive designs, we need not choose passivity instead. We may need to clarify in our own minds the difference between "offensive" and "initiative." An offensive is an assault, an attack, an attempt to seize control, the capacity to penetrate and devastate an adversary on his home ground. We are correct in challenging the ethics and efficacy of threatening such devastation. But in accepting this self-restraint, we should not also deny ourselves the freedom of action that is initiative. Indeed, without such initiative, creative and active rather than merely reactive and defensive, we will not likely succeed in defending ourselves.

Military strategists are correct in stating that no defense can succeed without some means of carrying the battle back to the adversary. Where they may well be wrong is in assuming the form of that offense to be an attack, executed with armed force. One can as well, and perhaps far better, assert one's freedom of action and take back the initiative by *non-military* methods, carrying the struggle back to the adversary by means of skilled diplomacy and trade.

The term "defense" itself tends to confine our consideration to only one side of the conflict, one dimension of the struggle. Defense as traditionally understood looks only to the safety of one's own, most often at the expense of one's adversaries. Defending oneself has traditionally been viewed as a kill-or-be-killed proposition in which all outcomes are either victory or defeat and only one can win. That understanding may have been appropriate enough, though fearfully destructive, in prior centuries, but the advent of nuclear weapons annuls victory and makes everyone a loser. Thus we must either stretch our consideration of the defense of the realm to *include* our adversaries or move to a naturally more inclusive concept that considers the fate of each party to the conflict and seeks to preserve *all* from harm. We extend this concern to our adversaries not from sympathy but from self-interest. We know well enough that if we do not look to their safety, we will have none ourselves. We can no longer gain security at their expense.

This is a subtle but significant step beyond defensive thinking and opens the possibility of developing strategies that transcend polarities instead of reinforcing them. It is also a step beyond relinquishing threats (a necessary and laudable move, but insufficient in itself) in that it does not simply block the illegitimate use of one's own armed forces but indicates an alternative direction for assertive and independent action. Mutually protective strategies must thus replace bids for unilateral advantage. The defense of one must *include* the defense of the other. What in the pre-nuclear age was a moral imperative has in the nuclear age become a practical necessity.

Yet this Olympian perspective may lie a ways above our present perch. We are still saddled with nationally based defenses composed of weapons which by their very definition and design are intended to provide security to only one side. So we may need to engage in a simultaneous process of relinquishment and redesign, giving up offensive weapons while inventing mutually preservative instruments and strategies. And we may need to reclaim defense from its illegitimate offspring, intervention and intimidation, as a transitional phase in moving our minds from unilateral advantage to common security.

"To provide for the common defense." The phrase is familiar to every American from the Preamble to the Constitution. When the words were first penned, the definition of "common" included exactly thirteen colonies on the eastern edge of the American continent. As the American empire expanded, so did the definition of the term "common." Yet there was always a frontier where the common defense ended and the common enemy approached. Even now, both the United States and the Soviet Union cling to the Bomb as their "common defense" against each other. Yet the self-evident truth of this era is that the one overwhelming threat both sides face is not each other but the shared peril of annihilation. It is in just this sense that the phrase "to provide for the common defense" takes on its most inclusive meaning: not "me" against "you," but both against the Bomb.

It may be that for those still more attached to national than global perspectives, expanding the definition of "common" to *include* their adversaries may supply the essential bridge between an obsolescent nationalism and a developing globalism. The world order thinking of peace researchers during the past few decades has yet to filter into the mainstream of political discourse. Seen from our imagined perch above the melee, we can readily devise a more workable plan for global security. But for the moment the actors are still national: nations still make history and global entities as yet do not. Thus we must find ways and means of translating new understandings of common security into the existing (and insufficient) paradigm of unilateral defense in order to overcome lingering resistances to new historical truths. "Think globally, act nationally," we might say, amending Rene Dubos' brilliant advice, "Think globally, act locally." Not because national actors are the best for the job but because they are for the moment the only ones with the effective power to make change. And even then, it is the common peoples of these nations who will need to force reluctant governments into providing for the common interest.

Theorists of alternative defense strategies are mostly quite aware of the limitations of perspective and possibility inherent in defensive thinking. It is generally seen as a strictly transitional strategy, a halfway house on the way to disarmament, a means by which to wean ourselves from our dependence on nuclear weapons without resorting to still other forms of destructive capability. Taken by itself, it may indeed constitute a technical fix, unlikely to be adopted and if adopted, likely to be harnessed to offensive weaponry in a new and more perverse Faustian bargain. But if adopted as one step among several or many, and if reinforced by more affirmative nonmilitary policies, alternative defense strategies could yet contribute to the emergence of a far less hostile environment than we now endure.

Questions

1. One man's defense is another man's threat. The offensive capability of a weapon depends as much on its location as on its range and firepower. To what extent would arsenals need to be refashioned to assure a thoroughly defensive capacity?

2. There is no perfectly invulnerable defense. Does a voluntary self-limitation renouncing the offense place the defense at an unrecoverable disadvantage? How can a defense without offense regain the strategic initiative?

3. What are the most likely threats to be defended against? The familiar scenario of a blitzkrieg may be wholly outdated. Invasion may not be nearly so likely as remote-controlled destruction and nuclear blackmail, against which both conventional military defenses and civilian resistance are largely helpless. How can these very different kinds of threats be addressed without resorting to nuclear counterthreats?

4. Can civilian resistance be united with a defensive military strategy without negating the strengths of each? What forms might this union take?

5. Some of the politicians and generals on whom a defensive strategy is being urged may not in fact desire just protection of their own territory. Offensive weapons provide them with the tempting capacity to extend their influence well beyond their own borders. Don't proposals for a defense without offense run contrary to their own immediate self-interest?

6. How would the Soviet Union and Warsaw Pact nations likely respond to NATO's adoption of a purely defensive posture? In what ways could the West encourage (or, if necessary, pressure) the East to respond in kind? What if they did not? Could public support for a defense-only policy be maintained in the absence of reciprocation?

7. Adoption of a defensive military arsenal in the absence of corresponding changes in foreign policy might well revert to an offensive stance over time. What political changes should accompany the transformation?

8. Is there any danger that in advocating a defensive military strategy these theorists will lose sight of the larger goal of disarmament? Is this another technical fix?

9. It may work for Switzerland, Sweden, and Yugoslavia, but can it work for a superpower with far-ranging interests and "responsibilities"?

10. What would prevent this strategy from degenerating into a new arms race in "defensive" weaponry, grafted onto an already uncontrollable competition in offensive arms?

Alternative Security: Not By Arms Alone

Closely related to the alternative defense tradition (some might say too closely for a distinction to be drawn) is the fledgling discipline of alternative security. While alternative defense theory generally confines itself to the hardware and strategies of military defense (except where it is blended with civilian resistance), an alternative security system includes a broad spectrum of non-military initiatives, which taken together help make the resort to armed force less likely in the first instance. By an integrated foreign policy of economic incentives, diplomatic initiatives, development assistance, participation in global institutions of governance, and the like, and by a domestic policy that emphasizes a robust but non-exploitive economy, an equitable distribution of wealth, and a vital cultural life, a well-conceived alternative security system creates an environment in which the problem of defense is no longer so overwhelming a preoccupation in international life. It is the systematic integration of these policies so that they reinforce one another which distinguishes the alternative security approach from the more random and piecemeal proposals of the arms control community and the exclusively hardware emphasis of alternative defense thinking.

Designs for alternative security systems vary in intellectual structure but share similar purposes and strategies. Robert Johansen lists five principles in what he calls a "global security system" serving the needs of citizens as well as states:

1. "It tries to prevent the desire for short-range advantages from dominating decisions at the expense of long-run interests."

2. "(It) emphasizes the importance of providing greatly expanded positive incentives rather than relying largely on negative military threats. . . . We can prevent any potentially aggressive government from precipitating a war if we can assure that, in the eyes of its governing officials, the benefits of peace outweigh the anticipated benefits of war."

3. "(It) emphasizes a positive image of peace which includes much more than war prevention—the right to peace and to freedom from the threat of genocide and ecocide; the right to security of person against arbitrary arrest, torture, or execution; the right to traditional civil and political liberties; and the right to fulfill all basic needs essential to life."

4. "It moves beyond the familiar, singular focus on security for one nation-state . . . . With a concern for the security of the whole nation and the whole human race, rather than merely for parts of either, a new attitude toward 'foreign' societies develops. The distinction between 'our' government and 'their' government begins to fade. . . . All governments become 'my' government."

5. "The most vital front line of defense becomes not a new generation of nuclear weapons but a new code of international conduct to restrict the use of military power."[16]

Uniting alternative defense strategies with a variety of non-military sources of security and strength, Johan Galtung identifies four elements in an effective alternative security system:

1. Transarmament from an offensive to a defensive military posture;

2. A gradual decoupling from superpower alliances;

3. Strengthening one's inner invulnerability (through cultivation of a sound and self-reliant economy, a healthy ecology, and an equitable and stable social system); and

4. Enhancing one's "usefulness" to would-be adversaries, by sponsoring mutually beneficial trade and providing valued services.

Regarding invulnerability, Galtung draws an insightful analogy with the health of the individual human body.

"The medical parallel is obvious: this is the capacity of the human body to withstand any insults . . . so that they do not have any 'bite.' The healthy body adequately nourished, clad, protected from the hazards of nature . . . is one important precondition. Another is the capacity of the body itself to fight off any intruder through the defense mechanisms of the cells of the human body. . . . And beyond that, should the other two fail, it is the capacity of the body to engage in a more lasting cure, to undertake its own repair work. Hence, what we are looking for would be the mechanisms by means of which the social body could do the same, all the time drawing on its own resources, not being dependent on an outside that might be partly or wholly hostile."[17]

Galtung suggests several dimensions of self-reliance to this "inner strength" he believes essential to a nation's security: economic self-sufficiency, especially in basic needs during times of crisis; ecological stability, nurturing the sources of food and shelter; political autonomy, preferably by means of a federal system; domestic social harmony, to avoid being weakened by internal conflicts; and cultural identity, a clear sense of purpose and shared values.

Approaching alternative security from the perspective of high technology in the space beyond the Earth's atmosphere, Daniel Deudney asserts that the thinking and planning of both mainstream and alternative strategists have yet to recognize the impact of what he calls "the transparency revolution" on the security of all nations. "Advances in information technology—sensors, communication and processing—have created a rudimentary planetary nervous system, fragments of a planetary cybernetic. . . . This transparency revolution means that the traditional struggle between offensive and defensive military force has been transformed into a competition between the visible and the hidden—between transparency and stealth. . . . Planetary-scale information systems bring the strategic competition between the superpowers to its least stable and most dangerous state. At the same time these systems make planetary-scale security possible for the first time in human history. Within the planetary war machine at its most advanced, unstable state may lie the embryo of a new security order."[18]

Deudney specifies four steps towards an alternative security system: "a new, more open information order, limits on weapons innovation, cooperative science and the pacification of the commons. . . . Although a new security system would require the regulation of the largest-scale systems of human creation, this could be done within a minimalist world order. A new security system need cover only those problems that are irreducibly planetary in scale. . . . "[19]

The Boston Nuclear Study Group, successor to the Boston Study Group, is a gathering of six researchers from the sociology departments of Brandeis and Boston College. They are collaborating on what began as a rebuttal to the Harvard study, *Living With Nuclear Weapons*, but has since evolved into a more comprehensive alternative to present American security policy. Their analysis begins with a critique of the "war chain" linking militarism, nuclearism, and security. It then introduces what the authors call a "survival politics"—new strategic and tactical doctrines, a panoply of unilateral initiatives, and use of "non-coercive modes of influence," which the authors believe may gradually supplant armed threats in international behavior. They envision a "community of nations" in which recourse to war is as unlikely as between the nuclear-armed members of the NATO alliance, and assert that certain trends and institutions already favor the transformation.

Earl Ravenal represents a school of thought uniquely his own in the otherwise generally congruent ensemble of alternative security systems currently being proposed. Once a Pentagon policy analyst, he now proposes a set of policies wholly contrary to the "power projection" interventionism that typifies present Pentagon thinking. Deeply skeptical of the prospects for establishing a "just" and stable world order (although not unsympathetic to its values), Ravenal predicts an increasingly anarchical political universe and counsels a policy of "strategic disengagement" to hedge against events that cannot be managed or controlled: "Others may be tempted to measure such an anarchical world, with its liabilities, against the abstract desirability of controlling events, steering them in favorable directions. But—somewhat like deterrence—attempts at control, if they fail, can yield both greater implication and greater harm for the nation that has made the attempt. Whereas, if we avoid involvement, there may be disorder in the world, but we are compensated by the fact that we are not implicated in it."[20]

Ravenal proposes a "new strategic paradigm" to govern U.S. policy in a world of "general unalignment":

"Instead of deterrence and alliance, we would pursue war-avoidance and self-reliance. Our security would depend more on our abstention from regional conflicts and, in the strategic nuclear dimension, on what I would call 'finite essential deterrence'. . . . Our military program would be designed to defend the most restricted perimeter required to protect our core values. Those core values are our political integrity and the safety of our citizens and their domestic property. . . . Over time, we would accommodate the dissolution of defensive commitments, including NATO, that obligate us contingently to overseas intervention."[21]

Ravenal's thinking has sometimes been dismissed as isolationist, but he maintains that a "strict and consistent" policy of military nonintervention can be well-mated with "a concern for constructive contact with the world. . . . The face of the policy would be hopeful and constructive, selectively accommodating interdependence—stressing practical cooperation in specific areas, encouraging the universal observance of an international law of self-restraint, and joining in the mediation of disputes and some limited but noncommittal peacekeeping. But the residual level of the policy would be skeptical and defensive, hedging and insulating against disorder."[22]

Finding Effective Lower-Risk Means of Defense

The Security Project of the World Policy Institute proposes a set of alternative military, foreign, and economic policies designed to cut $470 billion from the $1.720 trillion five-year projected Defense Department budget (leaving in place, of course, a still whopping $1.250 trillion). Its First Report proposes adoption of a "stable, streamlined defense":

"1) a nuclear deterrent force designed for the sole purpose of deterring nuclear attack. (6000 warheads based on 31 Poseidon and 6 Trident I submarines);

"2) the adoption of a non-threatening defense of Europe that would encourage detente between East and West, allow for the gradual withdrawal of American forces, and raise the threshold between peace and war;

"3) a program of non-intervention, political reconciliation, and economic development that would contain and reduce conflict in the Third World; and

"4) expanded support for the building of international institutions of common security for international peacekeeping, monitoring of arms, and the peaceful settlement of disputes."[23]

The report proposes a superpower "non-intervention regime," spearheaded by an independent U.S. pledge "not to send its armed forces to any country where they are not now present, even if invited."[24] Finally, in the realm of economic security, the report recommends a "full-employment, better-thy-neighbor" policy emphasizing the transfer of resources from military to "productive" activities, and a program of "public investment in the physical infrastructure and human capital" of the United States, to be called "USA-Works."

These sources constitute fragmentary evidence of the emergence of a new body of thought but it is too soon to say whether alternative security should be understood as a separate category from its close cousins, world order, alternative defense, arms control, and disarmament. Its focus is substantially broader than alternative defense and arms control but not so Olympian in perspective as world order studies often are. Alternative security thinking is an effort to shake the public mind free of its reflexive tendency to associate security solely with military hardware. By identifying and strengthening the very broadest range of non-military, "software" approaches to security, it seeks gradually to wean nations and peoples from their dependence on armed force, not by the direct approach of abolition but by the more indirect strategy of substitution. This perspective is founded on the premise that nations will not be persuaded to relinquish their weapons, nuclear or otherwise, until and unless a "better" (safer and/or more attractive) means of securing themselves can be found and made to work. The invention of alternative security systems is one effort to meet that need.

1. The European Security Study, *Strengthening Conventional Deterrence in Europe.* New York: St. Martins, 1983.
2. The Alternative Defence Commission, *Defence Without the Bomb.* London: Taylor and Francis, 1983.
3. Alexander Macleod, "Britain's Labour Party Spells Out Its Non-Nuclear Defense Plan." *Christian Science Monitor,* July 30, 1984, p. 27.
4. Frank Barnaby and Egbert Boeker, *Defence Without Offence: Non-Nuclear Defence for Europe.* London: Housmans, 1983, p. 58.
5. The Boston Study Group, *Winding Down: The Price of Defense.* San Francisco: W.H. Freeman, 1982, p. 11.
6. Harold Feiveson and Frank von Hippel, "Minimal Deterrence." Preliminary outline of a book, March 3, 1984, p. 1.
7. Randall Forsberg, "Confining the Military to Defense as a Route to Disarmament." *World Policy Journal,* Winter, 1984.
8. Jonathan Schell, "The Abolition," *The New Yorker,* January 9, 1984.
9. Freeman Dyson, *Disturbing the Universe.* New York: Harper and Row, 1979; and *Weapons and Hope.* New York: Harper and Row, 1984.
10. Dyson, *Weapons and Hope,* p. 277.
11. Ibid., p. 293.
12. Ibid., p. 290.
13. The term "transarmament" has been used in differing contexts by different scholars to describe a shift either from offensive to defensive weapons and strategies (Fischer and Galtung) or from armed force of any kind to civilian-based defense (Gene Sharp). The term apparently originated with Kenneth Boulding in the 1930s and was reintroduced by Theodor Ebert, the West German nonviolent defense theorist, in the 1960s. As used by Ebert, it referred to wholly nonmilitary defense. The confusion is a symptom, perhaps, of the lamentable lack of contact among peace researchers.
14. Dietrich Fischer, *Preventing War in the Nuclear Age.* Totowa, New Jersey: Rowman and Allanheld, 1984, p. 108.
15. Ibid., p. 113.
16. Robert C. Johansen, *Toward an Alternative Security System: Moving Beyond the Balance of Power in the Search for World Security.* World Policy Paper #24. New York: World Policy Institute, 1983, pp. 28-30.

17. Johan Galtung, *There Are Alternatives: Four Roads to Peace and Security*. Nottingham: Spokesman, 1984, p. 192.
18. Daniel Deudney, *Whole Earth Security: Towards a Geopolitics of Peace*. Worldwatch Paper 55, July, 1983, pp. 20-21.
19. Ibid., p. 41.
20. Earl Ravenal, *Defining Defense: The 1985 Military Budget*. Washington: The Cato Institute, 1984, p. 42.
21. Ibid., pp. 36-37.
22. Ibid., pp. 5, 14.
23. *The Security Project: The First Report*. New York: World Policy Institute, June, 1984, p. 23.
24. Ibid., p. 33.

It's a welcome sign that Harvard now has a Program on Nonviolent Sanctions in Conflict and Defense.* The director, Gene Sharp, can be introduced largely through the titles of his books. In 1973 he gave us *The Politics of Nonviolent Action;* in 1979, *Gandhi as a Political Strategist;* and in 1980, *Social Power and Political Freedom*—all of them published by Porter Sargent. In 1985 he added *Making Europe Unconquerable: The Potential of Civilian-based Deterrence and Defense* (Ballinger). The following paper is excerpted from "Making the Abolition of War a Realistic Goal," which won an award from the World Policy Institute.

Non-Military Means of Defense

Gene Sharp

NO BREAK IN THE CYCLE OF WAR IS POSSIBLE AS LONG as people and governments fail to perceive the existence and the effectiveness of alternative, non-military means of defense. Peace proposals and movements of the past have offered no credible alternative to military action. Their common failure, therefore, could have been predicted.

But the stubborn persistence of advocates of strong defense in considering only military means and failing to investigate non-military possibilities has led to the present dangerous situation and to the lack of development of possible options.

If we want to remove or reduce drastically the historic reliance on war and other violent conflict, it is necessary to substitute a "war without violence," by which people can defend liberty, their way of life, humanitarian principles, and their institutions and society at least as effectively against military attack as can military means.

Such a substitute defense policy would need to be one that can be held in reserve to encourage settlements without resort to open struggle. It would rest on the following premises.

***The program is at the Center for International Affairs, Harvard University, 1737 Cambridge St., Cambridge, MA 02138.**

Finding Effective Lower-Risk Means of Defense

The power of all rulers and governments is vulnerable, impermanent and dependent on sources in the society. Those sources can be identified: acceptance of the ruler's right to rule ("authority"), economic resources, manpower, military capacity, knowledge, skills, administration, police, prisons, courts and the like.

Each of these sources is closely related to, or directly dependent upon, the degree of cooperation, submission, obedience and assistance that the ruler is able to obtain from his subjects. These include both the general population and his paid "helpers" and agents. That dependence makes it possible, under certain circumstances, for the subjects to restrict or sever these sources of power by reducing or withdrawing their necessary cooperation and obedience.

If the withdrawal of acceptance, submission and help can be maintained in face of the ruler's punishments, then the end of the regime is in sight. Thus, all rulers are dependent for their positions and political power upon the cooperation of their subjects. This applies not only internally but also, with variations, in cases of attempted foreign invasion and occupation. The theory that power derives from violence, and that victory goes to the side with the greater capacity for violence, is false.

Instead, the will to defy and resist becomes extremely important. Adolf Hitler admitted that the problem of "ruling the people in the conquered regions" was "psychological": "One cannot rule by force alone. True, force is decisive, but it is equally important to have this psychological something which the animal trainer also needs to be master of his beast. They must be convinced that we are the victors."

The civilian population can refuse to be convinced.

A vast history exists of people who, refusing to be convinced that the apparent "powers that be" were omnipotent, defied and resisted powerful rulers, foreign conquerers, domestic tyrants, oppressive systems, internal usurpers and economic masters. Contrary to usual perceptions, these means of struggle by protest, non-cooperation and disruptive intervention have played major historical roles in all parts of the world, even in cases in which attention is usually concentrated on parallel or later political violence.

These unrefined forms of non-violent struggle have been used as major or predominant means of defense against foreign invaders or internal usurpers for the most part without preparation, training or planning.

These situations include: German strikes and political non-cooperation with the 1920 Kapp Putsch against the Weimar Republic; German government-sponsored non-cooperation in the Ruhr in 1923 with the French and Belgian occupation; major aspects of the Dutch anti-Nazi resistance, including several large strikes, 1940–1945; major aspects of the Danish resistance to the German occupation, including the 1944 Copenhagen general strike, 1940–1945; major parts of the Norwegian resistance to the Quisling regime and the occupation; and the Czechoslovak resistance to the Soviet invasion and occupation, 1963–1969.

The nature and accomplishments of the Czechoslovak defense are already forgotten by many and are being distorted when noted. The resistance ultimately failed, but it held off full Soviet control for eight months—from August to April—something that would have been impossible by military means. It also reportedly caused such morale problems among Russian troops that the first units had to be rotated out of the country in a few days and shipped, not to European Russia where they could report what was happening, but to Siberia.

All this was done without preparation and training, much less contingency planning. This suggests even in final defeat (as a result of capitulation by Czechoslovak officials, not defeated resistance) a power potential even greater than military means.

In addition to such cases as these, other resistance movements and revolutions against internal oppression and dictatorships are relevant: the 1980–1981 Polish workers' movement for an independent trade union and democratization; the 1944 revolutions in El Salvador and Guatemala against established military dictatorships; the 1978–1979 revolution against the Shah in Iran; the 1905–1906 and February 1917 revolutions in imperial Russia; the 1953 East German rising; the Polish movements of 1956, 1970–71 and 1976; the 1956–57 Hungarian revolution; the 1963 Buddhist campaign against the Ngo Dinh Diem regime in South Vietnam; the 1953 strike movement at Vorkuta and other prison camps in the Soviet Union; and diverse other cases.

This type of resistance and defense is possible against dictatorships, no matter how extreme, because they cannot free themselves from dependence upon the society they would rule. Dictatorships are not as strong and omnipotent as they would have us believe, but contain inherent weaknesses that contribute to their inefficiency, reduce the thoroughness of their controls, and limit their longevity. Those weaknesses can be located, and resistance can be concentrated at those cracks in the monolith. Non-violent resistance is much more suited to that task than is violence.

These experiences, to be sure, offer no ready-made substitute defense policy to be easily applied as a substitute for war. They do provide prototypes that could become the basis for one after analysis, refinement, planning, and training. Such a policy would be based not on military weapons and forces but on the civilian population and the society's institutions, its societal strength. An alternative to military defense is possible: "civilian-based defense."

This is a defense policy that utilizes prepared civilian struggle—non-violent action—to preserve the society's freedom and sovereignty. This is done not simply by efforts to alter the will of the attacker, but by the capacity to make effective domination and control impossible by massive and selective non-violent non-cooperation. It is possible to exert extreme pressure and even to coerce by non-violent means.

An attack for ideological and indoctrination purposes, for example, would likely involve non-cooperation and defiance by schools, newspapers, radio and television, churches, all levels of government, and the general population.

An attack aimed at economic exploitation would be met with economic resistance—boycotts, strikes, non-cooperation by experts, management, transport workers and officials—all aimed at reducing or reversing any economic gains of the attackers.

Coups d'etat and executive usurpations would be met with non-cooperation of civil servants, bureaucrats, government agencies, state and local government, police departments and virtually all the social institutions and general population to deny legitimacy and to prevent consolidation of effective control.

Police would refuse to locate and arrest patriotic resisters against the attacker.

Journalists and editors refusing to submit to censorship would publish newspapers illegally in large editions or many small editions—as happened in the Russian 1905 revolution and in several Nazi-occupied countries.

Free radio programs would continue from hidden transmitters—as happened in Czechoslovakia in 1968.

Clergymen would preach the duty to refuse help to the invader—as happened in the Netherlands under the Nazis.

Politicians, civil servants and judges, by ignoring or defying the enemy's illegal orders, would keep the normal machinery of government and the courts out of his control—as happened in the German resistance to the Kapp Putsch in 1920.

The judges would declare the invader's officials an illegal and unconstitutional body, continue to operate on the basis of pre-invasion laws and constitutions, and refuse to give moral support to the invader, even if they had to close the courts.

Teachers would refuse to introduce propaganda into the schools—as happened in Norway under the Nazis. Attempts to control schools could be met with refusal to change the school curriculum or to introduce the invader's propaganda, with explanations to the pupils of the issues at stake, continuation of regular education as long as possible, and, if necessary, closing the schools and holding private classes in homes.

Workers and managers would impede exploitation of the country by selective strikes, delays and obstructionism—as happened in the Ruhr in 1923.

Civilian-based defense operates not only on the principle that the price of liberty is eternal vigilance, but that defense of independence and freedom is the responsibility of every citizen.

This is a more total type of defense than the military system, since it involves the whole population and all its institutions. Because such participation must be voluntary in order to be reliable in crises, and because of reliance on non-violent means, however, civilian-based defense is intrinsically democratic.

As in military warfare, this type of struggle is applied in the face of violent enemy action. Casualties are, as in military struggle, to be expected. In this case, however, they are utilized to advance the cause of the defenders (by increasing their resistance) and to undermine the opponent's power (by alienating his own supporters).

There is no more reason to be dismayed by casualties, or to capitulate when they occur, than there is when they occur in military conflict. In fact, it appears that casualties in civilian struggles would be far lower than in military conflicts.

Civilian-based defense also has an attack capacity against usurpers and invaders—which one U.S. Army general has called "the sword of CBD." The basic dynamics of non-violent struggle would be aimed at undermining the will, loyalty and obedience of the attacker's troops, functionaries and administrators. The result could be to make them unreliable, less brutal in repression and at times mutinous on a large scale. This could dissolve the machinery of repression and administration.

Similar undermining efforts would be aimed at the enemy's usual supporters and home population, with the objective of producing dissent, disruption and opposition in his own camp.

Under some conditions, significant international opposition to the attack and support for the civilian defenders may be aroused. Occasionally this would involve international economic and political sanctions against the invader or internal usurper. These sanctions may be significant at times—witness the Arab oil embargo—although the defenders must primarily rely on their own actions.

Of the three broad theaters of defense—denial of the enemy's objectives, provocation of morale problems and unrest in the opponent's camp, and arousal of international support for the defenders—the direct blockage by the civilian defenders of the attacker's objectives is by far the most important.

Thoroughly prepared civilian-based defense policies in Western European countries would constitute a more powerful deterrent and defense against Soviet attack than do conventional weapons.

Although not without cost, civilian-based defense would be significantly less voracious in its consumption of the society's raw materials, industrial capacity, financial resources and energy supplies than is military defense.

Civilian-based defense would free the foreign policy of a country from controls based on the needs of its military policies. It would be conducive to development of foreign and international policies to assist the resolution of world problems, meet human needs and promote understanding and friendship.

Civilian-based defense could break the technological weaponry spiral and bypass the major problems of negotiated disarmament and arms-control agreements. With full recognition of international and domestic dangers, whole countries could mobilize effective capacities to prevent, deter and defend against attacks—while at the same time reducing and finally abandoning reliance on war.

7 Moving Toward the World We Want

War is an addiction. We act as if we could have "just one more" before we stop. But as a result of technological progress the habit's become too dangerous for the superpowers to indulge. Peruvian Indians chew coca leaves without apparent harm. When the leaves are refined into powdered cocaine, however, the results are very different. In military terms it's as if hunting spears had given way to machine guns. And when the powder is converted into "crack" to be smoked, the arms analogy would be strategic bombing. In a similar way, we have "refined" the arts of war until a contest has been converted into a catastrophe. Yet the theorists of "limited nuclear war" keep talking as if another little blast wouldn't be so bad.

Isn't it time to get off the stuff?

Any analogy breaks down at the edges, but if we continue to imagine war as a drug epidemic, then arms control would be similar to the partial "interdiction" of supplies. The U.S. government favors the solution of keeping out things it doesn't like. In Vietnam it tried to interdict the Ho Chi Minh Trail. Now it is working hard to stop the movement of illicit drugs across our borders. Both efforts, of course, have been failures. In the case of drugs, what if the government were to put more emphasis upon the demand side—upon the felt need for drugs in the kind of society that we have created? In the case of strategic arms, what if we were to focus less upon military hardware, as if it were being foisted on us, and more on the cycles of fear and provocation that lead both sides to acquire the weapons?

As a country we feel better, for a while, when we obtain a new kind of weapon. True, it's very expensive and it distracts us from other experiences of life. The satisfaction doesn't seem to last very long. But if one warhead per missile isn't enough, try three or maybe ten. If a submarine won't carry enough of the missiles, make it twice as big. If ballistic missiles don't do the trick, develop flying bombs that skim the earth's surface and contain digital maps of their exact flight path over the enemy's topography. If retaliatory threats get on the nerves, invent a defense scheme that will cost trillions, fill the heavens with ray-guns, and probably not work anyway. The process has the lunacy, the urgency, of an addiction.

But it feels so good. The arms race "creates jobs." (So does the drug traffic.) Weapons are technologically "sweet." (Crack makes you high, too.) Preparing for war is exciting. (So is dealing in illicit drugs.) Besides, how can we stop now?

I realize that this analogy will offend some readers. After all, we didn't freely choose the arms race—it was forced upon us, is forced upon us every day, by a ruthless opponent who, if we ever slacken our pace, will humiliate and maybe destroy us. But people doing "crack" may also feel driven by an irresistible force. At first it may be a feeling of emptiness, a desire for excitement. Then it's the need to continue with the drug. They don't rob people because they *want* to; they do it because they feel they *need* to.

The only way to stop is to stop.

In the case of arms and war, unfortunately, the situation is complicated by two parties needing to stop together. It's as if a couple were abusing a drug and, unless they got off it at the same time, one person would exploit, corrupt, or discourage the other. The most successful treatment programs seem to combine several elements—a confession by the client that he or she is out of control, a request for help, a pledge to stop abusing the substance, group support for each client by the others, and above all, development of "life-styles" that provide in other ways whatever the client was hoping to find in the drug.

The interesting question about drug epidemics in the U.S. is what aspects of our society allow drugs to appear so attractive, to those who abuse them, as compared with the available alternatives. Focusing upon the supply of drugs, as our government does, turns the attention on other people's greed—dealers in Latin America, smugglers, refiners—nasty types who would sell their own grandmothers or children if the price were right. In a similar way, the anti-war movement sometimes makes an enemy of the Pentagon, as others make an enemy of the Soviets.

But what about the demand side? What about the craving for

objects that will save us? On the part of both addict and narcotics officer, it's a fascination with the object, the white powder "with a street value of X million dollars," that keeps deflecting awareness away from the question of what else, other than cocaine, might make life worth "doing." In an analogous sense, we forget to ask in what other way, besides threats, we could relate to another great power.

Nobody will disarm into a vacuum. While becoming fully aware of the dangers of arms and seeking to limit those immediate dangers, we need to create alternative forms of security, to experiment with them, and to develop them more fully, before we will give up the weapons upon which we now depend. To fix our attention wholly upon the arms as such would distract us from the challenge of creating other ways of assuring our security.

One reason governments are so embarrassed by drugs is that widespread abuse of drugs is a vote of no confidence in the social reality created, tolerated, or evaded by governments. In "fighting" drugs, governments often engage in self-parody—that is, they display in exaggerated form some of the qualities that people may take drugs to get away from. In "protecting" us from enemies, they display some of the same qualities.

What qualities? When they can't think of anything wiser, governments like to "crack down." In this mood, governments like to forbid, punish, and retaliate—in general, to control. When a society comes to rely upon authoritarian threats, rulers fear that, if they ever relax their vigilance, all hell will break loose. Much of what we criticize in the Soviet regime are widespread governmental tendencies taken to an extreme.

Internally, the U.S. is far more relaxed. But when something goes wrong, such as the widespread abuse of drugs, we tend to try to control the symptom rather than heal the disease. This is not so different from Gorbachev's campaign against the abuse of vodka, which is said to reduce work efficiency, impair health, steal rubles from the family table, and (as *Pravda* keeps saying) reduce work efficiency. To the extent that the supply of vodka is reduced, we can expect to read about scandals involving other drugs—already a black market in opium poppies has arisen.

As Andrew Weil has reminded us, with the authority of a Harvard Medical School graduate, people like to get high. They always have. In addition to supporting and defending themselves, they like to expand their consciousness. Throughout history they have done it in many ways—through beating on drums, whirling, calling the name of God, ingesting parts of plants, making music, meditating, running, making love, speculating, fermenting grains, going on vision quests. In certain societies, activities such as these were not "counter-cultural"; they were among the major *gifts* of the culture. When expansiveness is integrated into a culture, instead of being suppressed or merely tolerated by it, that culture may experience less fear of going out of control.

What does all this have to do with U.S.-Soviet relations? Anybody looking at these relations for the first time would notice, above all, that they are characterized by fear and a desire to control by threats, with very little provision for expansiveness. We never get together with the Soviets and dream. We get together, intermittently, to negotiate about how to threaten one another in a slightly less risky manner. We complain about, and to, one another at length. We worry elaborately about the other side "getting ahead." But we have no provision for dreaming together, for sharing a vivid and detailed picture of what we desire rather than only what we fear.

If we *could* converse from the heart, what would we say? Even if there were only one chance in a thousand of going "beyond war," is it not worth considering the question of what we'd do there? A growing number of us are saying "Yes." It's not only because of the negative consequences of continuing on the present course; it's also because of the joys of expansiveness.

Of course there's a danger that, even by entertaining a hope, we become what Lenin, in his pungent way, called "useful idiots." He was talking about people in the West who were not tough-minded, who could be fooled by protestations of peaceful intent while a blitzkrieg was being mobilized behind the smiles. He is like the drug dealer who says to the new prospect, "Man, you're going to feel so *peaceful*."

We need always to be aware of deception, but we do not need

to adopt Lenin's view of human possibility, against the limits of which even the Soviets are bumping. Weapons need to be controlled, but at the same time, we can learn how to develop other methods for assuring our security. First we need to imagine these alternatives, then to project them powerfully and persistently as the basis of a new kind of relationship, and finally to break our habit of relying upon hits of weaponry.[1]

Before we can *get* what we want, we have to find the courage to dream up what it *is*, and to manifest that dream with the same energy and tenacity with which we are now maintaining our habit.

—Craig Comstock

1. For the account of how to imagine what we want and relate to adversaries in new ways, see *Citizen Summitry: Keeping the Peace When It Matters Too Much to be Left to Politicians* (Tarcher/St Martin's Press, 1986).

Moving Toward the World We Want

In this book, and elsewhere, many alternatives to the war system have been put forward. We know much of what needs to be done, or at least we know enough to begin. In political terms, however, how can we move toward the world we want?

One effort to move fast was the "nuclear freeze" campaign of the early 1980s. Like George Kennan's proposal for deep cuts, the freeze had the virtue of simplicity, and unlike deep cuts, it did not even require the military to give up any weapon that it already possessed. All the freeze required, as a first step, was that both sides would agree to stop building more. Emotionally, at least for TV viewers, the slogan recalled the moment in police dramas when an officer spots the suspect and yells, "Police! Freeze!"

The freeze was not only simple, dramatic, and restrained in its immediate ambitions; it also made use of a feature of American political life called popular voting. People were asked not merely to agree to something, or march in the street, or write to representatives. They were given the opportunity to vote yes or no for a proposition.

In this chapter, the rise and fall of the freeze campaign is analyzed briefly by Gordon Feller, a senior associate of Ark Communications Institute and author of a recent resource guide, *Taking the Next Steps,* and by Mark Sommer, who was introduced in the previous section. In their view, perhaps the major shortcoming of the freeze was its negativity. It said to stop the arms race; it did not say what to do instead.

Lessons of the Nuclear Freeze Initiative

Gordon Feller
Mark Sommer

IN ITS CURRENT EXPRESSION THE NUCLEAR FREEZE proposal ("a bilateral halt to the testing, production, and deployment of all nuclear weapons and delivery vehicles") was developed five or six years ago. Conceived by a group of associates for whom one highly determined woman, Randy Forsberg, became the principal formulator, the idea was framed as an initiative presented in public referenda throughout Western Massachusetts during 1981–82. It was, of course, not quite the habit of a New England town meeting to consider issues of global import. But in another sense it was altogether appropriate, since the earliest American colonies, like Swiss cantons even today, had used the town meeting as their essential means of self-governance. The New England town meeting has frequently been called the birthplace of American democracy.

Those voting in the blue highway villages of rural New England knew quite well that their opinions were of no great import to anyone other than themselves, but they had been enough disturbed by recent events that they decided to state their views in any case, having been asked by the initiative to consider them. First one, then another passed the freeze resolution, and by mid-1982, according to Randy Kehler, national freeze campaign coordinator at the time, more than 400 town meetings in the region had debated and passed the initiative. By autumn of that year it had been brought before voters in ten states and several dozen major cities across the country, comprising in all more than a third of the total U.S. electorate. The initiative had also passed the House of Representatives as a non-binding resolution. In addition, and perhaps more importantly, the United Nations, the Soviet Union, and a plethora of other citizen organizations and political bodies also endorsed the concept. In short, the freeze had been placed on the global political agenda.

The most striking aspect of the freeze phenomenon is that the idea came from altogether outside and beyond the boundaries of traditional political discourse. It was neither the lead campaign proposal for an aspiring politician nor the favored strategy of a political party. Far more adventurous a concept than any politician was yet willing to risk advocating on his own initiative, the freeze was a phenomenon in which politicians found themselves chasing after the bandwagon. They were only able to associate themselves with the idea because an awakened public had already cleared a space for it in the realm of mainstream political discourse.

But it was not just politicians who endorsed the concept. A second striking aspect of the freeze phenomenon was the epidemic of endorsements it triggered among groups one might never have expected to care. Doctors, lawyers, realtors, artists: for a time, each day seemed to produce a new and unlikely coalition, often as not constituted on the spot for the occasion of stating its views on the subject. The emergence of the freeze coincided with the rise of peace advocacy organizations in the professions which, clearly inspired by the example of the stout New England villagers and Helen Caldicott's energetic resuscitation of Physicians for Social Responsibility, asserted their right and authority to speak out, in their professional capacities, on issues of global import.

During the past year or more, the momentum of the freeze initiative has appeared to stall. Understanding why it so suddenly succeeded and then seemed so suddenly to fail will tell us much about the strengths and weaknesses of initiatives generated from the grassroots. It will also tell us of the power of the media in discouraging public engagement in the arms control process.

The freeze was, first of all, remarkably well-timed for success. Reagan Administration statements about the acceptability of limited nuclear war and its systematic preparations for "protracted global conflict" provoked such intense anxiety on both sides of the Atlantic that popular feeling positively ached for expression. In its startling simplicity, the freeze provided an apparently commonsense concept that even those unschooled in the intricacies of nuclear negotiation could readily grasp. It was like saying, "Stop it, already! Enough is enough!"

The tangible nature of the term, "freeze," provided a reference point in ordinary human experience for many who had never found anything remotely real about discussions of nuclear issues. Indeed, it was this very simplicity that critics ultimately held against the freeze: it's not simple, they said, but simplistic. Sophisticated designs for its implementation were published in the "serious" literature, but they never reached a general public already distracted by this largely groundless assertion, and the impression lingered.

In addition the freeze was, at least in its heady early days, a profoundly empowering experience for those who participated in one or another of its myriad campaigns. The organization, if so it could be called, was entirely decentralized and self-responsible; the main office was known as a "clearinghouse," as if to emphasize its non-executive nature. This dispersal of authority left local organizations free to establish their own ways and means of promoting the initiative and thus accorded them the respect due independent and self-reliant institutions. To draw comparisons with Poland's Solidarity movement would indeed be specious, given the immensely greater level of risk inherent in the Polish struggle. The freeze was a low-risk campaign carried on largely by middle-class professionals while Solidarity was an extraordinarily high-risk campaign among workers and dissident intellectuals. This difference reflects the vast contrast in social conditions between the two countries. But the two movements coincide in one essential respect: both represented a reawakening of political activism among people who had long assumed that their views held no weight in the political balance, and both groups were profoundly surprised when they found that they did.

But why has the freeze appeared to melt in recent months? One must first credit the adroitness of the Reagan Administration in derailing the momentum by a variety of shrewd diversions. Surely its most imaginative entertainment has been the selling of Star Wars as a "defensive measure," a superbly crafted inversion of the truth which, in addition, restakes the Administration's claim to the moral high ground. Election year produced a wave of distraction for at least twelve months preceding the election, a period during which debate on matters of substance is customarily suspended for the duration of the parade. In addition, the freeze may have been damaged by its own success. Americans are well-known to be susceptible to fads—this week the freeze, next week cabbage patch dolls. The passions through which we pass would make an incongruous and contradictory list. By the time the initiative reached a vote in the House of Representatives, the media had decided that the freeze had already "lost" the attention of most voters even though freeze coordinators successfully raised record numbers of citizen lobbyists to pressure politicians to approve it. Indeed, by that time the initiative had already been voted on in so many places and by so many people that a few benighted souls may even have assumed that the freeze had already been enacted.

Attuned to what looked like the ebbing attentions of their constituents, Congresspeople were able to vote a symbolic "yes" to a non-binding proposition against the nuclear peril while reserving a pocketful of "yes" votes for every particular weapon system the freeze sought to curtail. Freeze activists were well aware of the duplicity of this arrangement and insisted that individual Representatives would ultimately be held accountable by voters for contradictions in their voting records. But Americans are known generally to vote their pocketbooks before their principles, and while according to all polls, they continued to support the freeze by overwhelming margins, they ultimately voted in nearly equally overwhelming numbers for a politician promoting rearmament.

But the melting of the freeze at the public policy level cannot be attributed entirely to the ambivalence of the American voting public. Their perplexity reflects their twin concerns that nuclear weapons might kill them but that the Russians might otherwise do so first. While the freeze was intended to be mutual and bilateral, no one has yet come up with a formula the public believes will allow us to escape both fates. The public's reluctance to step into the vast unknown territory beyond nuclear deterrence demonstrates the incompleteness of the freeze as an alternative to the present system of international relations. The freeze says, "Stop right there! Don't move another inch!" It does not, however, say, "Go this way instead. It's much safer." Of course, this one initiative was never intended to answer all questions; it was intended, rather, as a way to begin asking them. But in the absence of a comprehensive design for an alternative structure of relations between East and West, this or any other single initiative seems more like a protest than a plan, a first step in a direction where we cannot see the next. Freeze theorists are fully aware of this problem and have been working hard to provide an itinerary for the journey of which the freeze is but the first stride.

How can an initiative strategy succeed? Our first lesson is that initiatives can only succeed if leaders and followers both lend their support. They will succeed up to a point when promoted by one group alone, at which point each in the absence of the other fails. So, for example, having endured enormous public pressure for agreement, President Kennedy and his new partner, Premier Khruschev, finally stepped into the political space opened in 1963 by popular support for significant arms control and signed a very limited test ban. But when domestic factional pressures began to impinge on the President's freedom of action, a larger public lobby was not there to exert a counterforce and provide the impetus for him to continue. Having received the initiative from "on high" (after having themselves produced the needed clamor), a great many people were grateful for what partial mercy had been rendered by the Limited Test Ban and retired from active service against the nuclear threat. They had accumulated no store of experience to counsel them to sustain the pressure long enough to obtain significant action.

In the case of the freeze, President Reagan placed his bets on the opposite strategy and effectively *froze* the freeze by deftly turning public attention to less burdensome and more alluring affairs—the invasion of Grenada and American "pride" in the L.A. Olympics—meanwhile proceeding oblivious to all suggestion. Whether in the past twenty years the public has developed greater stamina to remain with the campaign for the long haul is a matter still in doubt. So long as this President remains in office and the initiative remains with the executive branch, the freeze will remain in cold storage. (Frustrated by the obdurate resistance of the White House, the freeze campaign has recently developed a strategy focussing on a Congressionally mandated freeze.) But it is now on the political agenda of the entire planet, and is likely to remain there, waiting for a more favorable moment to be enacted, like human rights that have yet to be honored, like a world constitution still to be adopted, a promise still to be kept. Whether it *is* finally kept will depend on our will to persist to "the last full measure."

A second conclusion is that initiatives must be balanced in what they ask of the parties to the conflict, making equal demands of equal powers. Both the test ban treaty of 1963 and the freeze proposal were balanced in this respect, requiring identical conditions of both superpowers; neither could have gained public acceptance without that perceived balance. It would be difficult indeed for a president to sustain a policy of independent initiatives in the absence of essentially equal reciprocation. Any policy perceived to be unilateral or unbalanced will soon lose its majority. Balance is a highly charged and relative term in East-West relations, being largely a matter of perception. Precise comparisons, the bread and butter of traditional arms control, yield many disagreements and few accords. But the principle of parity is gradually coming to be accepted as the established relationship between East and West, and parity insists that both sides share equally the benefits and sacrifices of agreement. Any initiatives arising from independent sources must demonstrate this sense of basic fairness if they are to stand a chance of gaining broad public acceptance.

A third lesson we must learn is that individual initiatives should derive from and delineate a comprehensive design for an alternative system of peace and security to replace the war system in which we are now entrapped. Each piece in the puzzle should provide a hint of the design of the whole. This coherence and congruence implies a degree of sophistication and rigor in our thinking well beyond what we have yet achieved. Although libraries have been written about both arms control and disarmament, only passing references have been made to the nature and operation of an integrated peace system. The entire notion that peace can be organized as a system of mutually reinforcing processes and institutions is quite new to the global security debate and remains as yet unproven. Even the idea that war, in its profitable preparation and its bloody enactment, constitutes an integrated system is still a controversial proposition.

But the history of arms control makes amply clear that elements in the war system and the peace system are not simply interchangeable. As manipulated by superpower policymakers, arms control has been successfully appropriated by the war system. It no longer promises to terminate that system but simply to streamline and fine-tune it. A peace system would need to begin and always remain within the control of processes independent of the war system it seeks to replace. But in order to develop such a system, we must first conceive and imagine it in all its variegated complexity, a task of considerable magnitude in itself. Our inventions must be based not simply on abstract logic or wishful thinking but on the disenchanting lessons of actual experience. We must make a careful reading of history and a shrewd judgment of human motivations and the personalities of the parties in conflict.

And finally, initiatives should be, whenever possible, visible to and verifiable by a global viewing public. Initiatives should be framed in such a way as to require nations to perform tangible acts, acts that cannot easily be neglected or left undone. To most of the planet's residents, the arms race is a non-event: it can neither be watched nor heard, only heard about. Its invisibility effectively shields it from public scrutiny. One aim of the initiative process must be to give disarmament the high profile it deserves, to give us events we can actually witness: "disarmament by deeds instead of words." There is no reason why the theatrics of space shots and sporting events could not be transferred to the drama of a fully visible process of disarmament, though it might take a crisis to supply the essential passion. One could imagine, in the manner of Admiral Noel Gaylor's innovative "deep cuts" proposal, a series of warhead dismantling ceremonies, held on neutral ground between East and West (perhaps in Austria or Switzerland), open to inspection by all interested observers and televised (at least in the early events) to a much vaster global public. One could equally well imagine an individual superpower undertaking a series of public dismantling events, inviting friends, foes, and all between, and using the ceremonies as a means of pressing for tangible reciprocation by its adversary. Neither is imaginable at the moment, but crisis or concerted public pressure might sufficiently warm the atmosphere.

Nor should we necessarily assume that initiative strategies at the national level will remain in deep freeze for the duration of the present administration in Washington. In twin articles published in the Winter 1984 issue of *Foreign Affairs*, Kenneth Adelman, Director of the U.S. Arms Control and Disarmament Agency, and Paul Nitze, senior negotiator for the United States in the renewed Geneva arms talks, both appear to advocate an initiative strategy in place of arms control. The time may have come, writes Adelman, when "arms control without agreements" should be worked for. "In simple terms, each side would take measures that enhance strategic stability and reduce nuclear weapons in consultation with each other, but not necessarily in a formalized, signed agreement. Adopting this approach of individual, parallel restraint could help avoid endless problems over what programs to exclude, which to include, and how to verify them." Nitze argues that "Constructive discussion could lead to action on one side, concurrent with action on the other. Such an action-reaction process could conceivably provide a more solid foundation for progress in Soviet-American and East-West relations than a negotiated set of principles of conduct."

One hardly knows what to make of these pronouncements, arising as they do from the same administration that has given us Star Wars in the name of "defensive systems" and the MX, the "Peacemaker Missile," in the name of "bargaining chips." Is this an authentic strategy for arms reduction or a counterfeit gesture to disguise disinterest in arms negotiations? The point is still moot. But the shrewdest strategy for a global viewing public would no doubt be to take the Reagan Administration at its word and hold it to the promises of its rhetoric. If the Administration (and the Soviets) are indeed ready to undertake an initiative process, let us do all we can to help by providing them with a well-considered agenda and the essential popular pressure to make it stick.

Along with Harold Willens, another contributor to this book, Randall Forsberg was a leading figure in the nuclear freeze campaign. Like Mark Sommer, she urges that any military be truly *defensive*, and like the signers of the Five Continent Peace Initiative, she understands the need to propose a series of steps leading to a plausible, attractive future—not to focus attention, as the freeze did, solely on the first step.

Forsberg begins her analysis with the insight that military force is currently being used not only for defense, but on behalf of "intervention, repression, revolution, and deterrence." What if the major players agreed to give up each of these four activities? (In the case of repression, they might at least agree to stop doing it outside their own territory, or outside a clearly defined sphere of influence—which, in some cases, could more plainly be called a sphere of repression.) Restricting the military to defense would be revolutionary today, yet some armed forces have been dedicated to that function. Nobody fears an invasion by the Swiss, yet anybody planning to invade Switzerland would encounter tough resistance.

A Step-by-Step Approach

Randall Forsberg

THE NON-DEFENSIVE FUNCTIONS OF ARMED FORCE cannot be eliminated overnight. Instead, they must be eliminated through a series of clearly demarcated steps, each of which will create a plateau of military, political, and technological stability that can last, if need be, for decades. Each step must be durable and desirable in its own right, regardless of when or even whether any of the subsequent steps are taken. For before it can be demonstrated that any subsequent step will unquestionably increase stability and security, there must be a greater internalization of inhibitions against the use of force. Public debate that identifies and gradually overcomes opposition to the proposed next step is the primary means by which constraints on the accepted uses of armed forces can be internalized. And such debate may not reach a successful conclusion for years or even decades. Because people despair of ever achieving the ultimate goal of a disarmed peace, it would be extremely difficult to motivate widespread popular efforts for change without a set of powerfully attractive intermediate goals, each desirable in its own right. Thus a series of partial steps toward disarmament is needed to keep popular pressure for change high enough to prevail over the enormous psychological, economic, and political pressures to maintain the status quo.

In order both to strengthen inhibitions against the use of force and to win the support of the majority of the public, these gradual steps toward disarmament must eliminate the more provocative, aggressive, and escalatory functions of armed forces first, while leaving the more defensive and stabilizing elements intact. The history of the development of law and of attitudes toward violence in the domestic arena suggests that defense-oriented constraints on the use of force, far from being overly idealistic or excessively ambitious, are fully compatible with human nature. Moreover, the need for self-defense is the most fundamental human motivation for retaining military forces. Thus it should be easier, as well as ethically preferable, to abolish the less defensive functions of military forces first, the more defensive functions later. Because this process would parallel the historical development of legal and ethical constraints on the use of force within nations, domestic experience would provide a useful model for evaluating the potential uses of force in complex international conflicts. In the domestic arena, both internalized ethical standards and legal codes and procedures for the nonviolent resolution of complex conflicts are highly developed and widely recognized. Thus in international conflicts, in which parties naturally tend to see their own cause as just and that of their opponents as unjust, the degree to which a proposed use of force is truly defensive in nature can be judged objectively by the participants themselves, as well as by neutral arbiters, if we rely on analogies from domestic affairs.

Each of the steps outlined below is desirable in its own right. Each is designed to create a plateau of stability which can be maintained for an extended period of time. And each is aimed at eliminating the least defensive, most aggressive, escalatory, and provocative aspects of the military forces and policies that remain in existence at each stage of the process. In addition, each step is designed to encourage simultaneous change in the institutions of military policy and armed forces and in the attitudes that underlie the acceptable uses of force. These steps are not the only ones that would fulfill these requirements for achieving disarmament, but they do represent one plausible route.

The initial steps involve mainly the armed forces of the large powers: the United States, the Soviet Union, Britain, France, West Germany, China, and Japan. These few great powers account for nearly 70 percent of world military spending; more than 90 percent of the development and production of conventional weapons; and nearly 100 percent of the world's nuclear arsenals and nuclear-weapon production capacity. Thus the arms race can be reversed by stabilizing and then reducing the armed forces of these nations alone. This is fortunate, for, difficult as it may be to limit the war-making tendencies of a few big powers, that is a far easier task than resolving the many passionate and thorny Third World conflicts that motivate the maintenance of more than 100 smaller armies, many of them engaged in open war.

Step 1. Stop the production of U.S. and Soviet nuclear weapons and shut down their nuclear-weapon production facilities. This step constitutes a bilateral, U.S.-Soviet nuclear weapons freeze. By preventing the further development of dangerous preparations for nuclear warfighting, a bilateral nuclear freeze would preclude refinements in the ability of the superpowers to escalate from conventional war to nuclear war. Such refinements are not needed to deter out-of-the-blue nuclear attacks on cities, nor are they needed to deter a major conventional war in Europe. They are instead most closely associated with the least defensive contemporary use of conventional force: superpower military intervention in the Third World. A freeze on the production of nuclear weapons is the best way to begin changing the military system, for it would block those new weapons that are at once politically least defensive in their ultimate purpose and technically most threatening to human survival.

Step 2. End large-scale military intervention—the maintenance of military bases and the use of troops, air forces, or naval forces—by the industrialized countries of the northern hemisphere in the developing countries of the southern hemisphere. Such intervention represents the least defensive, most aggressive use of conventional military force in the modern world. Furthermore, each superpower fears that the other will undertake some intervention that will, sooner or later, pose a significant military threat to its own interests. Thus a sweeping halt to large-scale direct military intervention by the industrialized nations outside their perimeter—that is, in Latin America, Africa, the Middle East, South Asia, or the Far East—would allow developing nations far greater independence, sovereignty, and self-determination and would, at the same time, eliminate those uses of conventional force by either of the superpowers that the other finds most threatening.

Taken together, a nuclear freeze and a nonintervention regime would stabilize and defuse the most provocative aspects of the East-West military confrontation. Neither step involves any actual disarmament. Both not only would work in obvious ways to decrease the risk of a major nuclear or conventional war, but would do so without any subtle, offsetting increased risk of war. They would benefit not only the countries directly involved, but all countries. Yet even though they call for relatively modest changes in the existing military system, the first two steps may be the most difficult to bring about. For to accept the desirability of these steps is to admit that it is unwise, unsafe, and unnecessary to try to "manage" the existing military system indefinitely, as current military and arms control policies aim to do. Accepting the first two steps means admitting that we must begin to alter the military system in fundamental ways, so that, eventually, it can be largely or entirely eliminated. Thus the seeds of support for all the subsequent steps are contained in the affirmation of the first two.

Step 3. Cut by 50 percent the nuclear and conventional forces of the NATO and Warsaw Pact nations, plus those of China and Japan. By creating surplus tanks, aircraft, and ships, which would replace existing units when they wear out in training and exercises, this step would make it possible simultaneously to mothball the conventional-weapon industries and thereby to stop the wasteful and destabilizing technological race in conventional arms and halt the international trade in major conventional weapon systems.

By closing down the conventional-weapon industries, by ending technological advances in conventional weapons, and by stopping the international arms trade, this step would complete the process of stabilizing the major armed forces of the world. In addition, the 50 percent reductions in nuclear and conventional forces would begin the process of building confidence that war can be deterred with much smaller standing armed forces than exist today.

It is unlikely that we can take further steps, such as cutting much more than 50 percent of the nuclear and conventional forces of the industrialized countries, without first considerably changing certain nonmilitary conditions. This is particularly clear in Eastern Europe, where at least half of Soviet ground forces serve to maintain Soviet hegemony. Before the Soviet Union will be willing to eliminate most of the conventional forces now stationed in Eastern Europe, it will probably have to accept a wide range of civil liberties in that region. And for the Soviet government to permit wider civil liberties in Eastern Europe, it must probably allow them within its own borders as well. Furthermore, it probably will not be possible to persuade either superpower to eliminate more than half of its existing nuclear weapons as long as NATO and the Warsaw Pact retain conventional forces even half their current size. With five million men under arms on both sides, instead of the present ten million, the risk of a major conventional war would still appear too great for the nuclear arsenals to be reduced to, say, a few hundred strategic deterrent weapons on each side.

In other parts of the world, different but equally profound political and economic changes will be necessary before further substantial reductions of conventional armaments will be possible.

Step 4. Strengthen the economic development of Third World nations, promote civil liberties in all countries, and improve international institutions for negotiation and peacekeeping. These changes should, of course, be sought throughout the process of disarmament. But progress in these nonmilitary areas will be particularly important and feasible following the third step, which will release $300 billion per year from military spending and create tremendous political optimism and energy.

Between the third step and the fifth—each a clearly demarcated change—it might well be possible to make some further reductions in the standing armed forces of both industrialized and developing nations. Unprecedented military stability would have been created by a nonintervention regime, a halt to the technological arms race in nuclear and conventional weaponry, the end of the international trade in major weapon systems, and the 50 percent reductions in NATO and Warsaw Pact nuclear and conventional forces. Accompanied by some progress in nonmilitary areas, this stability might, for example, make it possible to reduce NATO and Warsaw Pact ground troops to one quarter of their present size and to trim down their navies and tactical air forces. Moreover, tactical nuclear weapons might be largely or entirely eliminated and strategic nuclear arsenals might be reduced to a few thousand nuclear weapons on each side.

Nonetheless, at this stage, the primary elements of the international system of armed deterrence would remain: the military alliances and mutual defense commitments; the superpower bases on foreign soil; and their long-range navies, tactical air forces, and strategic nuclear weapons. Even though these nuclear and conventional force structures would be far smaller than they are today, they would continue to help deter the outbreak of a conventional world war centered on Europe. They would continue to ensure, for example, that an internal disturbance in Eastern Europe would not lead to an all-out East-West war. They would continue to prevent the rearmament of a reunified Germany or the revolt of a chafing Eastern Europe. Only when the Soviet Union is willing to tolerate political independence and diversity in Eastern Europe, when Europeans are sufficiently secure in their demilitarization not to fear the rearmament of a reunified Germany, and when the United States is sufficiently sure that neither Europe nor the Soviet Union will rearm will it be possible to move on to the next step.

Step 5. Abolish all military alliances and foreign military bases and restructure conventional military forces to limit them to short-range border defense, air defense, and coastal defense. When the function of conventional military forces has been limited for some time to national defense, narrowly defined, so that major conventional war, both civil and international, has begun to be unthinkable, the next step can be taken.

Step 6. Abolish nuclear weapons. After all nations have had time to become sufficiently confident that armed force will not be used for any purpose other than national self-defense, and thus are sufficiently confident that force will never be used, it should be possible to take the final step.

Step 7. Eliminate national armed forces altogether and replace them with international peacekeeping forces. Whatever the merits of this particular path to disarmament, each of the steps listed above must be a central part of the process. Of course, for logical and practical reasons, certain steps must precede others. It is obvious, for example, that the production of nuclear weapons must be stopped before nuclear weapons can be abolished. Similarly, it is likely that superpower intervention in the Third World must be ended before U.S. and Soviet bases in Europe, which each considers far more vital to its security, can be eliminated. It is also clear that truly defensive military forces cannot be created until late in the process of arms reduction, but once military forces *are* truly defensive, it should be easy to demonstrate the stability of the system and to erase the fear of war. Only then will it be possible to think of complete global disarmament.

The proposals for a nuclear freeze and for noninterventionary conventional policies demonstrate that the most immediate obstacles to reductions in nuclear and conventional forces are not the relatively easy psychological and political obstacles associated with reducing purely defensive armed forces. The obstacles to eliminating forces still used for intervention, repression, revolution, and deterrence are far more difficult to overcome. Only when these non-defensive uses of force have been substantially or entirely eliminated will it be possible to reduce either nuclear or conventional forces substantially.

This proposed series of major, clearly demarcated steps constitutes not merely a safe, feasible route to disarmament, but also a checklist which those committed to peace can use to plan and evaluate coordinated campaigns. Each step provides a focus for action, a target for effort, and a standard to measure achievement along the way.

The defense-oriented approach to disarmament resembles that of pacifists and differs from that of arms controllers in that it is aimed, ultimately, at eliminating the military system. It differs from the pacifist approach, however, in that it concentrates on institutional change and condones the maintenance of defensive armed forces until a stable peace has been achieved. It shares with arms controllers the view that progress toward disarmament cannot be made in a single step, quickly, or unilaterally. It assumes that progress will require a series of steps taken over a period of decades through cooperative international action.

If a safe route to a stable disarmed peace is necessary and possible, as this article argues, then it is important to resume debate on disarmament, a topic neglected for twenty years. The ultimate goal of disarmament should be a stable peace established within the context of a resilient system of civil liberties and well-rounded, ecologically sensitive economic development. However difficult to achieve, this goal is of immeasurable value. By reversing the dangerous developments of the last four decades, it would ensure our survival—a minimal demand to make of our societies. But beyond ensuring mere survival, it would permit us to end the ancient, pernicious institution of warfare.

Distinguished as an economist, Kenneth Boulding has also been a leader in the field of peace studies and conflict resolution.* His "Seven Planks for a Platform" summarizes some of the main themes in this book, including "graduated reciprocation in tension-reduction" (Osgood, Etzioni, Willens), nonviolent responses (Sharp), "soldiers without enemies" (Channon, Keys, Sommer, Forsberg), strengthening world political organizations (Keys, Kurtz, Five Continent Peace Initiative, Forsberg), a U.N. "Spying Organization" (Kurtz, Clarke, Deudney), the role of "nongovernmental organizations" (Willens, Carlson, Rhodes, Greider, Norris), and research on peace (Sommer).

In weaving these themes together, Boulding incorporates many illustrations and asides drawn from his rich experience as an observer of governments and of anti-war movements. And he offers a model of the playful, inventive spirit necessary to think anything new, especially in the face of great danger.

Seven Planks for a Platform

Kenneth Boulding

WE ASSUME FAR TOO EASILY THAT EVERYTHING HAS TO be explicit. A richer and more realistic model of the social and international system would reveal the enormous importance of what is not said, not signed, but quietly taken as a rule of behavior. Without this element in social life, indeed, all societies would fall apart almost overnight. What I am looking for here is an almost half-conscious peace policy, communicated by nods, smiles, and raised eyebrows, and an atmosphere of intercourse which underlies the agreements, treaties, and declarations and often makes all the difference between failure and success.

The first plank in such a policy would be the removal of national boundaries from political agendas, except under circumstances of strong mutual agreement. A very large proportion of international war arises from dissatisfaction with existing boundaries and attempts to change them. . . .

***Kenneth E. Boulding, *Stable Peace* (University of Texas Press, 1978).**

A second aspect of peace policy would be the consistent pursuit of what Professor Charles Osgood has called "Graduated and Reciprocated Initiative in Tension-Reduction" (GRIT). . . . The process of détente between the United States and the Soviet Union, which goes back at least to Eisenhower and Krushchev and began perhaps with Khrushchev's doctrine of peaceful coexistence, has many aspects of the GRIT process. Its history indeed still remains to be written. These processes are by no means new and they go back a long way in human history, though they have not often been recognized as a species of process in the general field of the learning of peace.

The GRIT process begins by some rather specific, perhaps even dramatic, statement or act directed at a potential enemy (like Sadat's 1977 visit to Israel), intended to be reassuring and perhaps even naming or implying some act which might be taken in response, though this is somewhat dangerous if the suggested act is too specific. If the potential enemy responds, then a third act by the first party, a fourth by the second party, and so on could produce a dynamic of adjusting national images until the images become compatible. If there is an equilibrium to this process, it is to move toward compatibility of national images. And this, one suspects, can be spelled out in much greater detail than is usually done now. Most negotiation and interaction at the international systems level are conducted in an atmosphere of implicit national images, which can often be quite illusory. An exchange of information in these matters would play somewhat the same kind of role in the international system that the marriage counselor does in family conflict, where the communication system frequently produces mutually false images of the other party. In any conflict of two parties, A and B, there are at least four images involved—A's image of A, A's image of B, B's image of A, and B's image of B—and these can be very different. It by no means follows of course that, if A's image of B is the same as B's image of B, and B's image of A is the same as A's image of A, the conflict will be resolved. But, if these images are widely different, unnecessary and unreasonable conflict is likely to follow.

A peace dynamic is not enough, important as it is. At some point the dynamic must become explicit in a peace policy, expressed first perhaps in some dramatic public statement and act. This could be started by a single nation, preferably one of the superpowers. It could be embodied in a United Nations declaration or a treaty to which all nations would be invited to assent. The first step could be a public statement on the part of a major government, affirming the concept of stable peace and establishing it as the major goal of national policy. The main function of a statement or a manifesto is to create hypocrisy, which is a powerful agent of social change, for when actual policy is perceived to be too different from the professed statement a fulcrum for change toward the profession is provided. Manifestoes, however, tend to be ineffective unless they result in an organization. Neither the Declaration of Independence nor the Communist Manifesto would have come to much if they had not produced organizations to embody them and propagate them.

The first step in a peace policy after the manifesto, therefore, should be to set up a Department of Peace within the government with a number of missions. It should educate the public and the government in the meaning of stable peace and in the dynamics of peace policy through schools, the press, radio, television, publications, and so on. It should develop a research institute in the techniques of achieving stable peace, part of which might well be an in-house operation, patterned perhaps on the Stockholm International Peace Research Institute, and part of which might be a foundation, like the National Science Foundation, to encourage research in the universities and elsewhere. Part of this task should be the continuing collection and improvement of data on the conflictual aspects of the world system. Part of this would be a continuing program in the description and dynamics of national images, with a view to increasing their compatibility. There would also be a program like the present Arms Control and Disarmament Agency, one would hope on a greatly expanded scale.

A third element in peace policy would involve the serious exploration of both the theory and the practice of nonviolent responses to threats of violence, together with the formation of organizations to develop these nonviolent activities. Nonviolence is by no means a universal panacea, but it is one of the instruments of human betterment which deserves to be more clearly understood and prepared for, with regard to both its limits and its potentialities. There is now a large and serious literature on this subject. It can no longer be regarded as an eccentricity of saints. It should be part of the curriculum of every military academy. There should be institutes for its study and organizations to teach it. It represents an expansion of the agenda or repertory of the decision maker in many different types of social systems. It may be, as a high official in the Indian defense establishment once said to me, that it is much more suitable for aggression than for defense, more suitable for instance to create wanted social change than to defend against unwanted social change. This, however, needs to be investigated. What is most urgent is that nonviolence should be taken seriously.

A fourth aspect of a policy for peace would involve the piecemeal transformation of the role of the military in the direction of soldiers without enemies. This is already underway in such experiences as the United Nations forces in the Middle East, Cyprus, the Congo, and so on and in such organizations as General Rikhye's International Peace Academy. This is a real social invention. Its use so far has been very small compared with the enormous expenditures on soldiers with enemies and the national armed forces. Nevertheless, it represents a logical expansion of the role and the culture of the military in a direction that has enormous potential for the future. Whether existing military organizations can be transformed in this way may be doubted. The concept of the enemy is so fundamental to the military ethos, legitimacy, and morale that the new concept would undoubtedly be perceived as a threat to the existing military subcultures. It is significant, however, that much of the initiative in the development of the soldiers without enemies organizations comes from military men themselves such as Harbottle, who was the director of the United Nations forces in Cyprus. It may be that transformations of this kind would be welcomed by intelligent and sensitive members of the military who are disturbed by the fact that the traditional military ethos and culture have increasingly become a threat to human welfare in the altered state of military technology and the new conditions of the world. The destructive impact of the Vietnam War on the morale and legitimacy of the United States military establishment is the warning sign that the traditional military ethos may be on the point of collapse and that something new is needed.

A fifth plank in a peace policy would be national policies aimed at strengthening the structure of world political organizations, particularly intergovernmental organizations. The international nongovernmental organizations are only marginally within the purview of national policy, though there are problems with regard to permitting their operation within the nation. The great hole in the structure of world political organizations is the absence of any organization for negotiating disarmament. The United States and the Soviet Union have been making rather halfhearted attempts at SALT talks, but these have done nothing to reduce the general level of armaments or even reduce them at the points where they might cause the most trouble, namely at the borders. A United Nations Disarmament Organization, which would have the delightful acronym of UNDO, could act as a kind of marriage counselor in flitting back and forth among various decision makers, clarifying understandings, widening agendas, and removing obstacles to agreement. It could also play a role in policing, in inspecting agreements once they are obtained, and in monitoring the whole world war industry, as the Swedish International Peace Research Institute does now in a small way.

Bilateral disarmament negotiations are extraordinarily difficult in the absence of any mediating agency. It is not surprising that they have accomplished so little. There are many places in the world where partial and incomplete disarmaments, the moving of troops back from the frontier, inspection, and cooperation among armed forces could bring about both a lessening of the strain and an increase in the strength of the system. The effectiveness of a disarmament organization would obviously be determined partly by its own personnel, partly by the support given it by the various governments. That it is desperately needed can hardly be questioned. The almost inevitable proliferation of nuclear weapons and the grave potential instability of a multipolar system make the necessity for a United Nations Disarmament Organization all the greater.

I would also like to see a United Nations Spying Organization which would spy on everybody and publish the results immediately.

Secrecy in the international system is a very important cause in itself of strain. It produces extraordinary misapprehensions and illusions. Indeed, there is no segment of the social system in which the images of the world in the minds of the powerful decision makers are more deliberately formed by biased and inaccurate information. Even if one could not have a United Nations Spying Organization in the present state of mythology of the international system, one could at least have a United Nations Intelligence Organization for the study and publication of social indicators of strain on the system and, with more difficulty, its strength. With present techniques of the analysis of events data and content analysis, a great deal could be done to present the current picture of the international system as it moves from day to day in terms which are relatively free from bias and which are based on objective sampling. This one hopes would supplement and perhaps eventually supplant the vast apparatus of biased information collected and processed through spies, diplomats, state departments, and foreign offices.

Public intelligence organizations indeed are needed in many spheres of life to offset the inevitable corruption of information as it flows up through a large organization to the powerful decision makers. A friend of mine once visited a top official on the seventh floor of the State Department and found him sorting out the telegrams to go to the president, with only the favorable ones getting through. This seems to be repeated in every government and every large hierarchy in the world, and it is not surprising that so many bad decisions are made by powerful people. One could imagine what economic policy would be like if we relied for our economic information on spies collecting gossip in the chambers of commerce and labor union offices. This is a quite recognizable parody of how the so-called intelligence community works.

Perhaps one could even visualize a United Nations Organization for Image Transmission, with the delightful acronym of UNOIT. This would be concerned with the study and wide publication of the images which each country has of itself and of others and would deliberately seek to induce national governments to change images in the direction of compatibility. I doubt if such an organization would be very effective, but it might provide a symbol for a deliberate policy to change national interests. An important part of the theory of the international system is that national interest is a variable, not a constant, of the system. Changes in nations' images of their interests are of enormous importance in determining how well the system will work, so it is absurd to regard national interest as something fixed and almost God-given. Wide discussions created all around the world on what national interests should be could open up a new era of flexibility and a better possibility of these images of national interest moving toward compatibility through quite deliberate policies.

A sixth plank in a peace policy would be a policy on the part of governments toward nongovernmental organizations and the policies of these nongovernmental organizations themselves. The rise in nongovernmental organizations is one of the most striking phenomena of the last 150 years, beginning with the Universal Postal Union in the middle of the nineteenth century. There are now some five thousand of them. Many have a recognized status at the United Nations. The variety of the NGOs is of course immense. They range from scientific organizations to professional organizations, to church organizations, to peace organizations, and to political organizations. Because most of them are supported by some kind of grants economy or philanthropy, they tend to escape the kind of scrutiny and assessment which commercial organizations receive. They almost certainly add more to the strength of the world structure than they do to the strain, but this should not exempt them from examination. One would hardly expect the United Nations to have an office of philanthropy assessment, but a system of reporting and monitoring might be set up which would not be unduly restrictive and yet would help individuals decide which organizations to support. The competition of these organizations for membership and support is healthy if the information about them is reliable.

A seventh and last plank in a peace policy platform is, as might be expected, an emphasis on research. Peace research is a minuscule operation compared with the immense sums that go into military research and development. There are only a few hundred people engaged in anything that could be called peace research all around the world.

If this were substantially expanded, there is little doubt that some of the money would be wasted. One could hardly think of a better thing to waste money on, however, in light of the possible payoffs. I have been in the peace research movement myself for about twenty-five years. It has frequently been a discouraging and disheartening business, harder to finance I think than almost any other operation around a university. In part this reflects the strange limbo in which we find ourselves today, in that neither peace nor war is legitimate. The attitudes left over from a system of unstable peace make it hard for people to believe that concentrated, serious scientific endeavor could go into moving the world toward stable peace. The peace research movement itself is by no means a united enterprise. Nevertheless, over the years it has produced a substantial body of literature. Much of this is exploratory and tentative. We cannot claim any great successes like DNA or plate tectonics. It is absurd to suppose, however, that the social system in general and the international system in particular are not fit subjects for careful scientific inquiry. Our information system can be enormously improved and so can our theoretical structures. In light of the enormous urgency of the problem and the threat which war now represents to the continued existence of the human race, one would think that our strategy for the allocation of research resources would give peace research a high priority instead of the meager pittances it now gets.

These seven proposals are modest, they are all attainable within a reasonable amount of time, and they constitute a direction of change. Perhaps the greatest enemy of the human race is the very widespread feeling that its problems must be solved once and for all by some dramatic coup. This is not what the universe is like.

A professor at Harvard Law School, Roger Fisher clearly loves to persuade by telling stories. This is natural for a law professor in the Anglo-American system, who is accustomed to working from actual cases or, at least, from hypotheticals. In Fisher's case, the stories fall somewhere between the Stanley Kubrick of *Dr. Strangelove or How I Learned to Stop Worrying and Love the Bomb* and the figure of Nasrudin, celebrated in a thousand Sufi parables. One day Nasrudin goes to cash a check—it's an updated story—and when asked to produce identification he takes out a pocket mirror, looks at his face, and replies, "Yes, that's me all right!"

Who are *we* in the face of the bomb? In a Socratic way, Fisher keeps testing our assumptions and pushing us to see clearly, to take responsibility. And in the spirit of this book, he believes that "the way we can enlist support is less to burden others with guilt than to provide them with an opportunity to volunteer."

To Gain a Peace in the Nuclear Age

Roger Fisher

WHEN I SPOKE AT A CONFERENCE ON THE MEDICAL consequences of nuclear war last fall, several colleagues commented on my place at the end of the program. After two days on the horrors of nuclear Armageddon, I was supposed, in 45 minutes, to tell everyone how to prevent it. A typical remark was, "Boy, have *you* got a problem!"

Whenever I hear that phrase I am reminded of a small incident that occurred during World War II, when I was a B-17 weather reconnaissance officer. On one particularly fine day we were in Newfoundland test-flying a new engine. Our pilot was only a flight officer because he had been court-martialed so frequently for his wild activities, but he was highly skillful. He took us up to about 14,000 feet and then, to give the new engine a rigorous test, he stopped the other three and feathered their propellers into the wind. We flew along on one engine for a few minutes—it is rather impressive to see what a B-17 can do on one engine. It cannot quite hold its altitude, but if it is light, it can do quite well. Then, just for a lark, the pilot feathered the fourth propeller and turned off the engine.

That was startling. Suddenly the sound was gone. With all four propellers stationary, we glided, somewhat like a stone, toward the rocks and forests of Newfoundland. After a minute or so, the pilot pushed the button to unfeather. At this point he remembered that to unfeather the propeller, you had to have electric power, and to have electric power you had to have at least one engine going. As we were strapping on our parachutes, the copilot burst out laughing. Turning to the pilot he said, "Boy oh boy, have *you* got a problem!"

Though we're all in this together—like the crew of that B-17—I sense a tendency among professionals to put the problem of preventing nuclear war on someone else's agenda. But whoever is responsible for creating the danger, we're all on board one fragile spacecraft. We have no choice—we are here together and the risk is high. What can we do to reduce it?

We Have Met the Enemy

The risk is high for two kinds of reasons—hardware reasons and people reasons. We spend a lot of time thinking about the hardware. Both arms-control advocates and the military tend to focus on nuclear explosives and the means for their delivery. We think about the terrible numbers of terrible weapons, counting them by the hundreds, by the thousands, and by the tens of thousands. There are clearly too many, and there are too many fingers on the trigger.

Yes, changes should be made in the hardware. We should stop, I believe, all production of nuclear weapons and cut back on what we have. But even if we succeeded in stopping production and bringing about significant reductions, there would still be thousands of nuclear weapons for a long time. We, like the military, keep our attention on the hardware. They think it is the answer; we think it is the problem. It is not, however, the most serious part of the problem.

The U.S. Air Force and the U.S. Navy have a great many nuclear weapons. Each has enough weapons to blow the other up, and certainly they have disagreements—anyone who has stepped inside the Pentagon knows there are serious disputes between the air force and the navy. Disputes that mean jobs, careers—disputes that are sometimes more serious in practical consequences than disputes between the United States and the Soviet Union. Yet there is little risk of war between our air force and our navy. They have learned to conduct their disputes differently. They fight them out before the Senate Appropriations Committee and the Secretary of Defense, in the White House, and on the football field.

The case of the navy and the air force demonstrates, crudely, that the problem is not just in the hardware—the problem is in our own minds, in the way we think about nuclear weapons and in our working assumptions. And if the problem lies in the way we think, then that's where the answer lies. In Pogo's immortal phrase, "We have met the enemy and they are us."

Dangerous Assumptions

The danger of nuclear war is so great primarily because of the mental box we put ourselves in. We all have working assumptions that usually remain unexamined, and it is these assumptions that make the world so dangerous. Three that are particularly relevant concern our ultimate goals, the means for pursuing those ends, and finally, whose job it is to do what.

First, about our objectives. Internationally (as well as nationally and politically), we think we want to "win." Two stories illustrate that point. In England about fifteen years ago, my son and I were playing catch with a Frisbee in Hyde Park. Some Englishmen, who had apparently never seen a Frisbee, stopped to watch. Finally, one came over to me and said, "Sorry to bother you. Been watching you a quarter of an hour. Who's winning?" I wish I had been quick enough to ask him if he were married and if so, "Who's winning?"

The second story is about a time when I *was* trying to win. In the late 1950s I spent two years in the Solicitor General's Office arguing cases for the government in the Supreme Court. I started off with an excellent batting average—eight wins and no losses—which made me really impossible to put up with. Oscar Davis, who was then first assistant, put me in my place with some gibe which I naturally have forgotten. But he then said something I have always remembered. He said, "You know, we don't want to win them all." I said, "Excuse me?" He said, "Did you ever think what would happen if the government of the United States won all the cases in the Supreme Court? Prosecutors would run amok, respect for the Court would disappear, the whole concept of government under law would be destroyed—it would be a disaster." I said, "But Oscar, what am I doing up there? I put on my striped trousers and my morning coat, I go up, I argue. What is the purpose?" "Oh," he said, "we want to win each case, but not every case."

Internationally and domestically, we want a system in which we can play to win, but not in which any one side—even our own—wins all the time. But this concept of "winning"—that there is such a thing and that it is our dominant objective—is one of our fundamental beliefs. In fact, like a poker player, we have three kinds of objectives. One is to win the hand. Whatever we think we want, we want it now. We want victory. The second objective is to be in a good position for future hands. We want a reputation and chips on the table so that we can influence future events. We want power. Our third objective is not to have the table kicked over or the house burned down while we are playing. We want peace.

We want victory, we want power, and we want peace. And exploding nuclear weapons will not help us achieve any one of them. We have to reexamine rigorously our working assumption that in a future war we would want to "win." What do we mean by "win"?

The Meaning of Success

Last year, I taught some exercises in Rome for the NATO Defense College. I gave the officers a hypothetical war in Europe and asked them to work out NATO's war aims. The war was presumed to have grown out of a general strike in East Germany, with Soviet and West German tanks fighting on both sides of the border. Deterrence had failed. I told the NATO officers in Rome, "You are in charge of the hotline message to Moscow. What is the purpose of this war? What are you trying to do?" At first they thought they knew—win! Very simple. But what did that mean?

As time passed, they realized that NATO did not plan to conquer the Soviet Union acre by acre as the Allies conquered Germany in World War II—they did not plan physically to impose their will on the Soviet Union. They were seeking a Soviet decision; that was the only way they could have a successful outcome.

With further thought they reached a second conclusion—they were not going to ask for unconditional surrender. That gave the NATO officers a simple task: just define the Soviet decision that would constitute success for NATO and that NATO could reasonably expect the Soviet Union to make. They worked through the day considering how the Russians saw their choice.

It turned out that the only plausible objective was to stop the war. "Winning" meant ending the war on acceptable terms. It was with difficulty and even pain that some officers discovered that winning meant stopping, even if there were some Soviet troops in West Germany and only a promise to restore the *status quo ante*.

They found it hard to draft a fair cease-fire that didn't sound like a unilateral NATO ultimatum. It might say, "Stop firing at 0100 hours tomorrow, promise to withdraw, promise to restore the status quo within 48 hours, and we then will meet in Vienna to talk about serious problems." But the NATO officers did not know whether the Soviets would prefer Geneva to Vienna or whether they wanted 0200 hours rather than 0100 hours, and so forth.

Someone creatively suggested, "Wouldn't it be a good idea if we worked out ahead of time with the Russians some cease-fire drafts that we can both accept?" Another officer was incredulous: "What did you say? You are going to negotiate the armistice before the war begins? In that case, why have the war?" Good question.

The need to reexamine our assumptions about the purpose of our foreign policy is also demonstrated by our self-centered definition of national security. Typically, political leaders suggest that the first priority of foreign policy is national security, and only after that has been taken care of should we worry about our relations with the Soviet Union and China. Such thinking assumes that somehow we can be safe while the Soviet Union faces a high risk of nuclear war. But in any nuclear war with them, missiles will go both ways. There is no way we can make the world more dangerous for them without also making it more dangerous for ourselves. The less secure the Soviets feel, the more they will do something about it. Security is a joint problem.

We should say, "Look, you Russians have got to understand why we build these missiles and how it looks to us when you behave as you do. You must take responsibility for helping us deal with our security problem." Similarly, we must take responsibility for dealing with their security problem.

I may point out that those of us who say such things frequently do not practice what we preach. I am always prepared to tell my friends at the Pentagon that it does no good to call the Soviets idiots. "Don't you see, you idiot?" I say. We in the peace movement frequently think our job is to "win" the war against hawks. In worrying about our interests, we often assume that our adversaries have none worth considering. That's wrong. We have to find out what their legitimate concerns are, and we have to solve their legitimate interests in order to solve our own. At every level, domestically and internationally, we need to reexamine our objectives. We are not seeking to win a war but to gain a peace.

Plutonium Security Blanket

A second set of dangerous assumptions are those we make about means—about how to pursue our objectives. Again, we and the rest of the world are caught up in a military way of thinking.

The basic mistaken assumption is that for every problem there is a military solution. We will first try diplomacy, but if that doesn't solve the problem, we assume that we can always use force. We recognize that military means may cost a lot. But most of us believe that if we have the courage and are prepared to pay the price, then we can always solve the problem by military means. The reality, however, is that the big problems in today's world have no military solution. Nuclear war is not a solution—it is worse than any problem it might "solve."

We have mislearned from the past. During World War II, the Allies could physically impose a result on Hitler acre by acre. But the world has changed; we can no longer physically impose results on any nuclear nation. Except for imposing modest results in small situations,. the only means we have is to try to change somebody's mind. There is no way we can make the world work by using nuclear bombs, and yet people put that in the back of their minds. "Yes, that's true," they say, and then they go right ahead operating on the assumption that there are military solutions.

Like Linus in the Charlie Brown comic strip, we cling to our security blanket, military hardware. Both American and Soviet officials clutch their blankets as though they offered real security. Somehow, we think this bomb, this hardware, will protect us and let us avoid dealing with the real world.

Seeing Things as the Other Side Sees Them

We all know far better ways to deal with international problems: break up big problems into manageable pieces. Look at each item on its merits. Sit down with the other side and discuss it. Don't concentrate on what people say their positions are but try to understand and deal with their interests. Communicate and, particularly, *listen.* What's bothering them? Before we can change their minds, we have to put ourselves in their shoes. How would we feel?

If you were sitting in Moscow and saw Japan thinking of rearming; if you saw your long-time strongest ally, China, with a 2000mile common frontier, now becoming your worst enemy; if you saw Pakistan apparently building a nuclear bomb; if you heard Western voices saying, "We must help the rebels in Afghanistan"; if you saw American military equipment in the Persian Gulf, in Saudi Arabia, in Egypt, and in Israel; if you saw Greece going back into NATO and Turkey in the hands of a military government; and if cruise missiles were about to be located in West Germany—how must all that look from Moscow? Is there any reason for the Russians to be a little concerned? The only way we can reduce the threat of war is to affect their future thinking, and the starting point is to understand their present thinking.

Second, we have to invent wise solutions, find good ways to reconcile our differing interests. And both sides must participate in that process. It's not possible in any conflict for one side to produce the right answer. The understanding that comes from working jointly on a problem, and the acceptability that comes from participating in creating a solution, both make any answer better.

We will want to insist on principles in the problem-solving process—objective criteria independent of their will and ours. The best method cannot be to insist, "We're more stubborn than you." That way lies chaos, that way lies Armageddon. We can be firm on principle and flexible in application, and at every point, participation, participation, participation.

The same process applies equally to our domestic differences. Again, we in the peace movement are not the only source of wisdom—we are part of the conflict. There are a lot of people in this country who have legitimate concerns about the Soviet Union. We have got to put ourselves in their shoes—Pentagon shoes. We should not insist on inventing all the answers ourselves. We must try to participate with them, not carry on a war.

We will need to promote joint problem solving, not just at the intellectual level but at the level of feeling, the level of emotion, the level of caring, the level of concern. International conflict is too often dealt with cerebrally, as though it were a hypothetical problem. We need to apply what we know, and even more, to keep on learning about human behavior. We want to understand how to affect it—not merely to manipulate it, but to realign the forces within us to work in a better direction.

Whose Job Is It?

The danger of nuclear war also comes from my third set of assumptions—about whose job it is to reduce the risk of war. If there were a military solution, there would be a case for leaving it to the military, to policy-science experts and professional strategists. At the meeting last fall on the consequences of nuclear war, doctors were saying, in effect, "We are just concerned with the medical aspects. We will limit ourselves to the area of our professional expertise. We will tell you how bad a nuclear war would be. It is somebody else's job to prevent it." Such statements rest on the assumption that preventing war is in the military hardware department. But we are not facing a technical military problem. The solution lies within each of us, in changing our assumptions—and in changing other people's assumptions. The solution lies in reaching maturity, in abandoning our plutonium security blanket.

We are dealing with a set of psychological problems with dangerous physical consequences. If people are clinging to a plutonium blanket that is bad for their health, you do not get an engineer to design a better blanket—the problem is within those who are clinging to it.

No one has a professional license in reducing the risk of nuclear war. Fortunately, however, no professional license is required. But who has the skills to deal with psychological problems such as denial and turning flesh-and-blood issues into jargon-laden abstractions? Who is likely to notice that people are denying responsibility because a problem seems too overwhelming? Weapons engineers? I think not.

A little while ago I left you in a B-17 over the hills of Newfoundland—our copilot was telling the pilot that *he* had a problem. Well, we didn't crash; we weren't killed. On that plane we had a buck sergeant who remembered that behind the bomb bay, we kept a putt-putt generator in case we landed at some emergency air field that did not have electrical power to start the engines. He ran back, fiddled with the carburetor, wrapped a rope around the flywheel a few times, pulled it and pulled it, got the generator going, and before we were down to 3000 feet we had electricity. Saving that plane was not the sergeant's job, in that the danger was neither his fault nor his responsibility. But it *was* his job in the sense that he had an opportunity to do something about it.

We professionals tend to define our roles narrowly. In teaching law students, I ask, "What would have been the responsibility of professional musicians hearing Nero's performance on the fiddle while Rome burned? Should they limit themselves to discussing the music?" A citizen's response would presumably be to get a bucket and help put out the fire. By becoming professionals, do we become less responsible? Can we say, "No, I'm a professional, not a firefighter. It's someone else's job to put out the fire"?

Our special knowledge and training may not obligate us to try to prevent nuclear war, but it does give us an opportunity. My notion of who has responsibility for something is best defined by who has the opportunity—we have the opportunity. I encourage you, as I encourage myself, to use it. The world is at risk. The very danger of nuclear war means that there is more opportunity for each of us to make a difference than ever before in human history.

A Professional's Responsibility

What are some of these opportunities? If everyone with any significant power made the right decision every time, that would be as close to utopia as we could get. There are only three reasons people don't make the right decisions. One is that they are bad decision makers. Our job is to find better ones—that is what politics is about. Second, they are operating on bad assumptions. Our job is to correct their assumptions. And third, they are subject to constraints such as misguided public opinion. Our job is to remove those constraints.

To get a wise decision, we need to tackle every aspect of a problem in every way possible. No amount of useful research will overcome poor decision makers, no number of good decision makers will overcome bad assumptions or harmful constraints. Somebody has to propose a solution. Somebody has to put it on a decision maker's agenda. Somebody has to persuade others that it is a good idea, and somebody has to carry out the idea. There is enough to keep us all busy.

We need hard facts. We need theories on how to reduce instability. We need to develop knowledge about nuclear war, about the consequences, and about ways to reduce those risks. We need to communicate that knowledge, both to the public, which constrains our decision makers, and to the people who are making important decisions. We need to communicate both the bad news and opportunities for reducing it. When we deliver bad news and then say there is nothing to be done about it, the bad news does not become operational. The bad news about nuclear war will not by itself reduce the risk of nuclear war; we have to act on that news.

Blood on the White House Carpet

My earliest arms-control proposal dealt with the president's remoteness from reality when facing a decision about nuclear war. A young officer follows the president with a black attaché case containing the codes needed to fire nuclear weapons. I envisioned the president at a staff meeting considering nuclear war as an abstract option. He might conclude, "On SIOP Plan One, the decision is affirmative. Communicate the alpha line XYZ." Such jargon holds reality at a distance.

My suggestion was quite simple. I proposed to place the code number in a little capsule that would then be implanted right next to the heart of a volunteer. The volunteer would carry a big, heavy butcher knife as he or she accompanied the president. If ever the president wanted to fire nuclear weapons, he would have to kill one human being personally and realize what an innocent death is. Blood on the White House carpet—it's reality brought home. When I suggested this to friends in the Pentagon they said, "My God, that's terrible. Having to kill someone would distort his judgment. The president might never push the button."

A Happy Venture

Whether or not this particular idea has any merit, a lot of action is required to educate the public. A common lament is, "I don't know what to do." That gives you something to do right there. Get a half-dozen friends together some Saturday morning and figure out some things you might do. Identify three or four other people whom you think could make decisions of some significance. What are some of these decisions? Why haven't they made them already? What can be done to increase the chance they'll make a desired decision next week? Whoever it is—journalists, people in government, business people, civic organizations, professional societies, a friend of President Reagan's—what are some things they might do that would illuminate our faulty working assumptions or help establish better ones? More time should be spent on aiming than on implementation. In intellectual efforts, as in gunnery, one's aim is crucial.

Don't wait to be instructed—take charge. This is not an organized campaign that someone else is going to run. If you share these concerns, get involved—there is a lot to do. Perhaps you are still holding onto your own security blanket, that nice professional definition of your job. The security blanket most of us cling to is, "Don't blame me. It's not my job to plan nuclear strategy. I'm not responsible for the risk of nuclear war." The first thing you can do is give up that security blanket.

The way we can enlist support is less to burden others with guilt than to provide them with an opportunity to volunteer. I find it a happy venture. It is a glorious world outside. The sun is shining. There are people to love and pleasures to share. Details of past wars and the threat of the future should not take away the fun and the joy we can have working together on a challenging task. I see no reason to be gloomy about trying to save the world—about working together to reduce that risk by 1 or 2 percent. There is more exhilaration, more challenge, more zest in tilting at windmills than in any routine job. Be involved, not just intellectually but emotionally. Here is a chance to work together with affection, with caring, with feeling. Let some of your emotions hang out. Don't be so uptight. You don't have to be simply a scientist, a lawyer, or a business person. We are human beings—be human.

People have struggled all their lives to clear ten acres of ground or simply to maintain themselves and their family. Look at the opportunity we have. Few people in history have been given such a chance to apply their convictions, their values, their highest moral goals with such competence as our professional skills may give us. A chance to work with others—to have the satisfaction that comes from playing a role, however small, in a constructive enterprise. It's not compulsory; so much the better. But what challenge could be greater? We have an opportunity to improve the chance of human survival.

Acknowledgements

It was Don Carlson who, despite the pressures of an intense business life, noticed that most "peace books" have actually been antiwar books, often dependent upon arousing fear rather than encouraging hope. He asked the crucial question, "What's being done or proposed that's productive but not yet widely known or fully employed?" He imagined a positive book that would include and stimulate images of a peaceful world. While Carlson provided financial support to the project through the Ark Communications Institute, his more remarkable contribution has been asking the right questions. He also came up with the idea of running quotations alongside the text of *Citizen Summitry*, gathered hundreds of brief passages from his reading, and worked closely with me in the final editing.

It was my role to develop this approach into the particulars of a book project by creating an editorial structure that would express our vision, exploring available material (both published and newly written) and selecting the best of it, commissioning other papers, finding people to help with the project, editing the texts, and writing the introductions to the various sections and chapters.

What started as a single book grew and eventually became two—*Citizen Summitry* and *Securing Our Planet*. Of this pair, the first deals with communication, people-to-people exchanges, self-transformation, and the envisioning of alternative futures; the second, with particular steps that can be taken to create a less dangerous, more satisfying world. Both volumes are published in the U.S. by Jeremy P. Tarcher, Inc. of Los Angeles and distributed to the book trade by St. Martin's Press of New York.

In editing the two books, Carlson and I received advice from friends, colleagues, and experts too numerous to list, including many of the contributors and others active in the field, whether in government, academic life, business, the media, or public interest organizations. To give only a single example, both of us have benefited by involvement in Business Executives for National Security (BENS), a bipartisan organization with an appeal to conservatives disturbed by Federal deficits attributable to military spending, liberals worried about the arms race that is sustained by that spending, and military reformers who believe there's nothing more dangerous than a "defense" establishment that's not only costly and provocative but arguably ineffective.*

For advice and specific leads, Carlson and I turned above all to Gordon Feller, an Ark colleague who is well-informed about the field. (Some of his knowledge is now available in *Taking the Next Steps*, a succinct guide to resources, described below in "Other Ark Products.") In obtaining the Soviet contribution to *Citizen Summitry*, we had the assistance of Joel Schatz, a citizen diplomat who created an experimental computer network between the U.S. and the Soviet Union. (Schatz has a chapter in *Citizen Summitry*, and Feller in *Securing Our Planet*.) At the start of this project Elaine Ratner assisted with research, and at the end Matt Chanoff edited the introductions and Mary Dresser helped with the flow of material. Linda Lazarre made a number of contributions, including the Ark logo. Dick Schuettge supervised production. Type was set by Classic Typography of Ukiah, California, under the eye of Stan Shoptaugh. Quickly grasping our intent in editing these books, Jeremy Tarcher offered his full support in publishing them.

Thanks to all of you for your commitment and your articulate energy.

—Craig Comstock

*See Gary Hart (with William S. Lind), *America Can Win: The Case for Military Reform* (Adler & Adler, 1986).

The editors gratefully acknowledge the following publishers and authors for permission to reprint the following material:

SETTING POSITIVE MUTUAL EXAMPLES

"The Kennedy Experiment" by Amitai Etzioni: excerpted from "The Kennedy Experiment," *Western Political Quarterly*, Spring 1967. Reprinted by permission of *Western Political Quarterly*.

"The Next Initiatives" by Harold Willens: from pp. 102–109 in *The Trimtab Factor*. © 1984 by Harold Willens. By permission of William Morrow & Company.

"Getting to Yes" by Roger Fisher and William Ury: from chapter 1 of *Getting to Yes* by Roger Fisher and William Ury. © by Roger Fisher and William Ury. Reprinted by permission of Houghton Mifflin Company.

REDEFINING THE ROLE OF BUSINESS

"The Trimtab Factor" by Harold Willens: from pp. 26–27 in *The Trimtab Factor*. © 1984 by Harold Willens. By permission of William Morrow & Company.

"Being Dead Is Bad for Business" by Lucien Rhodes: from *Inc. Magazine*, July 1984. By permission of the publisher.

SHIFTING TO A GLOBAL PEACETIME ECONOMY

"The Economics of Security" by William Greider: excerpted from "Does a Bear Sit in the Woods?" *Rolling Stone*, January 17, 1985. © 1985 by William Greider. By permission of the publisher.

"The Impact of Trade on Soviet Society" by Cyril E. Black and Robbin F. Laird: this essay originally appeared in *Common Sense in U.S.-Soviet Trade*, edited by Margaret Chapman and Carl Marcy, published by the American Committee on East-West Accord, Washington, D.C., 1983. By permission of the publisher.

COOPERATING IN OUTER SPACE

"The President's Choice" by McGeorge Bundy, George F. Kennan, Robert S. McNamara, and Gerard Smith: from *Foreign Affairs*, Winter 1984. Reprinted by permission of *Foreign Affairs*. © 1984 by the Council on Foreign Relations, Inc.

"Star Wars and the State of Our Souls" by Patricia M. Mische: from *Star Wars and the State of Our Souls: Deciding the Future of Planet Earth*. © 1984, 1985 by Patricia M. Mische. Published by Winston Press, Minneapolis, Minnesota. All rights reserved. Used with permission.

"Speech in Search of a Statesman Capable of Delivering It" by Howard Kurtz: originally titled "A Proposed Speech for the President of the United States," prepared by War Control Planners, Inc. By permission of War Control Planners, Inc.

"The Age of Transparency" by Kevin Sanders: originally published as "Etiquette for the Age of Transparency," *Whole Earth Review*. By permission of the author.

"Unlocking Space" by Daniel Deudney: from *Foreign Policy*, Winter 1983–84. Reprinted with permission from *Foreign Policy*. © 1983 by the Carnegie Endowment for International Peace.

FINDING EFFECTIVE LOWER-RISK MEANS OF DEFENSE

"Deep Cuts" by George F. Kennan: published as "A Proposal for International Disarmament" in *The Nuclear Delusion: Soviet-American Relations in the Atomic Age* by George F. Kennan. © 1982 by George F. Kennan. Reprinted by permission of Pantheon Books, a division of Random House, Inc.

"I Cut, You Choose" by Stephen H. Salter: originally published as "Stopping the Arms Race: A Modest Proposal," *Issues in Science and Technology*, Winter 1986. © 1986. By permission of *Issues in Science and Technology*, the National Academy of Sciences, 2101 Constitution Avenue, Washington, D.C. 20418.

"David and Goliath" by Freeman Dyson: Chapter 4 of *Weapons and Hope* by Freeman Dyson. © 1984 by Freeman Dyson. Reprinted by permission of Harper & Row, Publishers, Inc.

"Alternative Security Systems" by Mark Sommer: from pages 3–37 of *Beyond the Bomb: Living Without Nuclear Weapons* (Expro Press/The Talman Company, 1985).

"Non-military Means of Defense" by Gene Sharp: from Gene Sharp, "Making the Abolition of War a Realistic Goal" (New York, World Policy Institute, 1981, as abridged in the *Des Moines Sunday Register* of 17 January 1982). For information on related publications, write to: Publications Director, Program on Nonviolent Sanctions, Center for International Affairs, Harvard University, 1737 Cambridge St., Cambridge, MA 02138.

MOVING TOWARD THE WORLD WE WANT

"Lessons of the Nuclear Freeze Initiative" by Gordon Feller and Mark Sommer: from *Syntropy: A Quarterly Journal*, Summer 1986. (*Syntropy: A Quarterly Journal*, 48 Shattuck Sq. #73, Berkeley, CA 94704.) By permission of the authors.

"A Step-by-Step Approach" by Randall Forsberg: excerpted from "Confining the Military to Defense as a Route to Disarmament," *World Policy Journal*, Winter 1984. Reprinted by permission of the author and the Institute for Defense and Disarmament Studies.

"Seven Planks for a Platform" by Kenneth E. Boulding: from *Stable Peace* by Kenneth E. Boulding, © 1978 by University of Texas Press. By permission of the author.

"To Gain a Peace in the Nuclear Age" by Roger Fisher: from *Technology Review*, April 1981. Reprinted with permission from *Technology Review* (a publication of the M.I.T. Alumni Association). © 1981.

Listed in the order of the Table of Contents, the following chapters were written for *Securing Our Planet* or appear here for the first time as a whole or in this form:

"Breaking the Trance" by Don Carlson

"The Way GRIT Works" by Charles Osgood

"Proposal for a World Peace Corps" by Robert Fuller

"The First Earth Battalion" by Jim Channon

"Peace-Keeping Forces" by Donald Keys

"Eisenhower's Legacy" by Thomas J. Watson, Jr.

"Technology and Freedom" by Regis McKenna

"Security and the Bottom Line" by Don Carlson

"Joint Ventures with the Eastern Bloc" by William C. Norris

"Speech in Search of a Statesman Capable of Delivering It," by Howard Kurtz

"Alternative Security Systems," by Mark Sommer

The following chapters are drawn from public papers or other material not covered by copyright to the best of our knowledge:

"A Strategy of Peace" by John F. Kennedy

"The Moral Equivalent of War" by William James

"All the Skills Necessary" by John F. Kennedy

"Farewell Address" by Dwight D. Eisenhower

"Everybody's Satellite," public testimony by Arthur C. Clarke

"The Delhi Declaration" by the Five Continent Peace Initiative

creating an ark as big as the earth

A Word About Ark

When the flood came, Noah was able to save a remnant of life by taking it aboard his ark. Today, facing the possibility of nuclear war, we need to create an imaginary ark as big as the earth itself. Safety lies not in trying to find a refuge, but only in preventing a general catastrophe. In order to avoid war, we need to become global citizens—not in place of our loyalties to an organization, an ethnic group, a religion, or a homeland, but in addition to them. To the extent that we build in our minds an ark to encompass the earth, we can not only prevent disaster but enlarge the possibilities of life.

In this spirit, Don Carlson, a successful entrepreneur, founded Ark Communications Institute in 1985 as a non-profit operating foundation. Ark shares the goal of some "anti-war" groups, but differs in at least two ways from most of them. While respecting the need to oppose evils, we put most of our energy into developing positive alternatives. And while treasuring much of what is usually meant by "peace," we accept conflict as part of being alive, believing that, like anything else, it can be handled destructively or creatively. We see the energies that are sometimes expressed in conflict as a resource, like fire, to be employed. Society does not try to "ban" fire or put it out wherever it's found; we put it to work in furnaces, stoves, boilers, and engines.

In dealing with conflict between two sides, Ark tries to look for common ground, not in the sense of a lowest common denominator, but of a third way, perhaps in the form of a shared task. While aware of the prolonged attempts to deal with many destructive conflicts, we also often feel that almost nothing has been tried.

As Einstein once observed, many problems can be dealt with only by shifting to a new level. On this new level, people can often communicate in ways that were formerly blocked. To us, communication has many senses. We have begun using high technology to link people in the two superpowers and to link Americans working for a better world. In addition to books, we are interested in using mass media to show visions of enemies becoming adversaries—a word that President Kennedy was careful to use—adversaries becoming rivals, rivals becoming competitors, and competitors possibly becoming partners.

We are developing methods to help people think and communicate on more levels than they ordinarily do. We take a deep interest in the connection between inner growth and political change, believing that neither can reach its full potential without the other. For this reason, we are exploring methods for each of us to find inner peace by getting in touch with neglected or formerly inaccessible parts of ourselves.

Our research thus ranges from inner development to international relations. One day we are creating a "game" in which widely diverse and quarreling participants imagine a positive future; on the next, investigating alpha wave biofeedback equipment, and on the third day, planning a trip for citizen diplomats to the Soviet Union.

In our first year, Ark has sponsored this book and a companion volume, edited by two of the principals in the institute; helped to create a computer link and the potential for slow-scan TV space-bridges between the U.S. and the Soviet Union; produced a series of radio documentaries starting with "Beyond the Boundaries" and "A Better Game Than War"; served as creative consultants for a major TV show; set up meetings between government officials and influential citizens in the area of arms control; consulted with think-tanks on public policy; helped to create "Peace Net," a data base and electronic mail system; and sent our staff to speak at a wide variety of conferences and forums.

As a non-profit operating foundation, we sponsor our own programs rather than giving grants to others. These programs include products such as this book, the sale of which helps sustain our broader work. An affiliated entity, the Ark Foundation, makes small grants to related programs conducted by outside organizations and individuals.

One way to keep in touch with the range of activities in which we take an interest is through *Global Partners*, our newsletter. To receive a free sample copy, please send us a stamped business-size envelope addressed to you, or simply request a copy when ordering any of Ark's products described below. Our address is Ark Communications Institute, 250 Lafayette Circle, Lafayette, CA 94549.

Other Ark Products

In addition to this book, *Securing Our Planet* (price, $10.95), the following items are available through the Ark Communications Institute:

1. *Citizen Summitry: Keeping the Peace When It Matters Too Much to be Left to Politicians*, a companion to this volume, also edited by Don Carlson and Craig Comstock and published in the U.S. by Jeremy P. Tarcher, Inc./St. Martin's Press, 1986 (price, $10.95). A complete table of contents for this book appears below.* Order CITIZEN SUMMITRY.

2. A 60-minute audio cassette containing two professionally produced half-hour radio programs, "Beyond the Boundaries" and "A Better Game Than War," both of which consist of further thoughts from contributors to *Citizen Summitry* and *Securing Our Planet* (price of the cassette together with a transcript of it, $9.95). Order AUDIO CASSETTE.

3. *Taking the Next Steps*, by Gordon Feller: a succinct guide to the most useful resources for readers who want to explore further (and to act upon) ideas contained in the Ark Communications Institute books (price, $2.95). Order TAKING NEXT STEPS.

4. *Peace Trek*, a "family coloring book," 42 pages in large format, answering the question, "What would the world look like if peace broke out?" (price, $5.95). Details from the art appear in *Citizen Summitry* as illustrations to the chapter by Elise Boulding. *Peace Trek* has been called "interesting, beautiful and effective" (Linus Pauling, Nobel Laureate), and "a wonderful tool to help families explore practical options for waging peace" (Gerald G. Jampolsky, M.D.). Order PEACE TREK.

Securing Our Planet and *Citizen Summitry* are available from your bookseller, and *Peace Trek* from selected stores that sell items for children. Or you may order any of the books, as well as the audio cassette, direct from:

Ark Communications Institute
250 Lafayette Circle
Lafayette, California 94549

Please specify the items you want and the quantity for each, print your address, and enclose a check or money order for the total price (with appropriate sales tax for California residents), plus $1 per order for handling. Allow 4–8 weeks for delivery.

*Table of contents for the companion volume:

Citizen Summitry: